I0012022

Mathematica Data Analysis

Learn and explore the fundamentals of data analysis with the power of Mathematica

Sergiy Suchok

BIRMINGHAM - MUMBAI

Mathematica Data Analysis

Copyright © 2015 Packt Publishing

All rights reserved. No part of this book may be reproduced, stored in a retrieval system, or transmitted in any form or by any means, without the prior written permission of the publisher, except in the case of brief quotations embedded in critical articles or reviews.

Every effort has been made in the preparation of this book to ensure the accuracy of the information presented. However, the information contained in this book is sold without warranty, either express or implied. Neither the author, nor Packt Publishing, and its dealers and distributors will be held liable for any damages caused or alleged to be caused directly or indirectly by this book.

Packt Publishing has endeavored to provide trademark information about all of the companies and products mentioned in this book by the appropriate use of capitals. However, Packt Publishing cannot guarantee the accuracy of this information.

First published: December 2015

Production reference: 1151215

Published by Packt Publishing Ltd.
Livery Place
35 Livery Street
Birmingham B3 2PB, UK.

ISBN 978-1-78588-493-1

www.packtpub.com

Credits

Author
Sergiy Suchok

Reviewer
Shivranjan P Kolvankar

Commissioning Editor
Amarabha Banerjee

Acquisition Editor
Manish Nainani

Content Development Editor
Sumeet Sawant

Technical Editor
Vivek Arora

Copy Editor
Kausambhi Majumdar

Project Coordinator
Dinesh Rathe

Proofreader
Safis Editing

Indexer
Rekha Nair

Graphics
Jason Monteiro

Production Coordinator
Aparna Bhagat

Cover Work
Aparna Bhagat

About the Author

Sergiy Suchok graduated in 2004 with honors from the Faculty of Cybernetics, Taras Shevchenko National University of Kyiv (Ukraine), and since then, he has a keen interest in information technology. He is currently working in the banking sector and has a PhD in Economics. Sergiy is the coauthor of more than 45 articles and has participated in more than 20 scientific and practical conferences devoted to economic and mathematical modeling.

About the Reviewer

Shivranjan P Kolvankar is a teacher and a passionate embedded system developer.

He did his masters in instrumentation science from the University of Pune in 2014. He has worked on statistical process control charts and data analysis for his masters' thesis.

He has experience in working with Bluetooth low energy, embedded system development, C#.NET, VB.NET, and Android application development.

Currently, he is working with Teach for India as a teacher with underprivileged and low income kids. He believes that quality education that caters to the learning ability of a child is their fundamental right.

He applies a head-heart-hand strategy to teach mathematics.

When he is free, he loves to play the flute and tinker with Arduino and Sensor Interfacing.

www.PacktPub.com

Support files, eBooks, discount offers, and more

For support files and downloads related to your book, please visit www.PacktPub.com.

Did you know that Packt offers eBook versions of every book published, with PDF and ePub files available? You can upgrade to the eBook version at www.PacktPub.com and as a print book customer, you are entitled to a discount on the eBook copy. Get in touch with us at service@packtpub.com for more details.

At www.PacktPub.com, you can also read a collection of free technical articles, sign up for a range of free newsletters and receive exclusive discounts and offers on Packt books and eBooks.

https://www2.packtpub.com/books/subscription/packtlib

Do you need instant solutions to your IT questions? PacktLib is Packt's online digital book library. Here, you can search, access, and read Packt's entire library of books.

Why subscribe?

- Fully searchable across every book published by Packt
- Copy and paste, print, and bookmark content
- On demand and accessible via a web browser

Free access for Packt account holders

If you have an account with Packt at www.PacktPub.com, you can use this to access PacktLib today and view 9 entirely free books. Simply use your login credentials for immediate access.

Table of Contents

Preface

There are many algorithms for data analysis, and it's not always possible to quickly choose the best one for each case. The implementation of algorithms takes a lot of time. With the help of Mathematica, you can quickly get a result using a particular method because this system contains almost all known-algorithms for data analysis. If you are not a programmer but you need to analyze data, this book will show the capabilities of Mathematica that use just a few strings of intelligible code to solve huge tasks ranging from statistical issues to pattern recognition. If you're a programmer, with the help of this book, you will learn how to use the library of algorithms implemented in Mathematica in your programs, as well as how to write algorithm testing procedures.

With each chapter, you'll immerse yourself more into the special world of Mathematica. Along with intuitive queries for data processing, the nuances and features of this system will be highlighted allowing you to build effective analysis systems.

What this book covers

Chapter 1, *First Steps in Data Analysis*, describes how to install the Wolfram Mathematica software and starts us off by giving a tour of the Mathematica language features and the basic components of the system: front end and kernel.

Chapter 2, *Broad Capabilities for Data Import*, examines the basic functions that are used to import data into Mathematica. You will also learn how to cast these data into a form that is convenient for analysis and check it for errors and completeness.

Chapter 3, *Create an Interface for an External Program*, focuses on the basic skills to transfer accumulated data-processing tools to Mathematica, as well as to use Mathematica's capabilities in computing expressions in other systems.

Chapter 4, Analyzing Data with the Help of Mathematica, covers Mathematica's functions that help to perform data classification and data clustering. You will know how to recognize faces, classify objects in a picture, and work with textual information by identifying the language of the text and recognizing it.

Chapter 5, Discovering the Advanced Capabilities of Time Series, profiles the various ways to process and generate time series. You will find out how time series processes are analyzed and become familiar with the main model type of these processes such as MA, AR, ARMA, and SARIMA. You will able to check observation data for stationary, autocorrelation, and invertibility.

Chapter 6, Statistical Hypothesis Testing in Two Clicks, deals with hypothesis testing on possible parameters. Several examples are provided, which will check the degree of dependence of data samples and test the hypothesis on true distribution of the samples.

Chapter 7, Predicting the Dataset Behavior, takes a moment to look at some useful functions that help in finding regularities and predict the behavior of numeric data. We'll take a look at the possibilities of intelligent processing of graphical information and even imitate an author's style expanding their work or restoring it. Using the methodology of probability automaton modeling, we will be able to build a model of a complex system in order to make predictions with the parameters of the system.

Chapter 8, Rock-Paper-Scissors – Intelligent Processing of the Datasets, tackles the creation of interactive forms to present research results. Also, Markov chains are considered with functions that help in finding the transition probability matrix. In the end, we will cover how to export results to a file for cross-platform presentations.

What you need for this book

To follow the examples in this book, you will need a copy of Wolfram Mathematica 10.1 or higher. If you wish to follow one of these examples and you do not have the right edition, you can download the trial version from the Wolfram website.

For some examples, you will need to use Windows 7 as your operating system but it is not necessary.

Who this book is for

The book is designed for those who want to learn how to use the power of Mathematica to analyze and process data. Perhaps you are already familiar with data analysis, but have never used Mathematica, or you know Mathematica, but you are new to data analysis. With the help of this book, you will be able to quickly catch the key points for a successful start. If you perform data analysis professionally and have your own algorithms written in any programming language, you will learn how to optimize computations by combining your libraries with the Mathematica kernel with the help of this book.

Conventions

In this book, you will find a number of styles of text that distinguish between different kinds of information. Here are some examples of these styles, and an explanation of their meaning.

Code words in text, database table names, folder names, filenames, file extensions, pathnames, dummy URLs, user input, and Twitter handles are shown as follows: "In order to import data into Mathematica, the Import function is used."

A block of code is set as follows:

```
int main(int argc, char *argv[]) {
  return WSMain(argc, argv);
}
```

Any command-line input or output is written as follows:

```
checkdate[x_] :=
If[DateObject[x] == DateObject["2015JAN"], True, False]
```

New terms and **important words** are shown in bold. Words that you see on the screen, in menus or dialog boxes for example, appear in the text like this: "To go to further instructions, let's click on the **Continue** button".

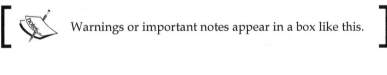

> Warnings or important notes appear in a box like this.

> Tips and tricks appear like this.

Reader feedback

Feedback from our readers is always welcome. Let us know what you think about this book—what you liked or disliked. Reader feedback is important for us as it helps us develop titles that you will really get the most out of.

To send us general feedback, simply e-mail feedback@packtpub.com, and mention the book's title in the subject of your message.

If there is a topic that you have expertise in and you are interested in either writing or contributing to a book, see our author guide at www.packtpub.com/authors.

Customer support

Now that you are the proud owner of a Packt book, we have a number of things to help you to get the most from your purchase.

Downloading the example code

You can download the example code files from your account at http://www.packtpub.com for all the Packt Publishing books you have purchased. If you purchased this book elsewhere, you can visit http://www.packtpub.com/support and register to have the files e-mailed directly to you.

Errata

Although we have taken every care to ensure the accuracy of our content, mistakes do happen. If you find a mistake in one of our books—maybe a mistake in the text or the code—we would be grateful if you could report this to us. By doing so, you can save other readers from frustration and help us improve subsequent versions of this book. If you find any errata, please report them by visiting http://www.packtpub.com/submit-errata, selecting your book, clicking on the **Errata Submission Form** link, and entering the details of your errata. Once your errata are verified, your submission will be accepted and the errata will be uploaded to our website or added to any list of existing errata under the Errata section of that title.

To view the previously submitted errata, go to https://www.packtpub.com/books/content/support and enter the name of the book in the search field. The required information will appear under the **Errata** section.

Piracy

Piracy of copyrighted material on the Internet is an ongoing problem across all media. At Packt, we take the protection of our copyright and licenses very seriously. If you come across any illegal copies of our works in any form on the Internet, please provide us with the location address or website name immediately so that we can pursue a remedy.

Please contact us at `copyright@packtpub.com` with a link to the suspected pirated material.

We appreciate your help in protecting our authors and our ability to bring you valuable content.

Questions

If you have a problem with any aspect of this book, you can contact us at `questions@packtpub.com`, and we will do our best to address the problem.

1
First Steps in Data Analysis

Wolfram Mathematica is not just a system that can solve almost any mathematical task. It is a *Swiss army knife*, which allows you to process images, audio, and written text to perform classification and identification of objects, as well as to manage files. It also provides other tremendous capabilities. In Mathematica, you can call functions, as well as receive a call from modules that are written using different programming languages and technologies, such as C, .NET, Java, and others. The system can generate code and compile files. Mathematica includes over 4,500 functions, which go beyond mathematical calculations. Wolfram Mathematica is constantly changing while still maintaining full compatibility with previous versions.

In this chapter, you will learn the following:

- How to install the Wolfram Mathematica system
- How to configure it to fit your needs
- What the Wolfram language kernel is and how to run it
- Wolfram Mathematica features to write expressions

System installation

Depending on your needs, you can work with Mathematica both online and offline. For a deeper understanding of the system and its settings and capabilities, we will use the offline version. In this book, we'll deal with examples of Mathematica, version 10.2 (the current version at the time of writing this book), which are fully compatible with all future versions.

You can install Mathematica on the major operating systems: Windows, Linux, and Mac OS X.

To download a trial version of Mathematica, use this link: `https://www.wolfram.com/mathematica/trial`. You will be prompted to create a new Wolfram Mathematica ID or sign in with your already registered ID:

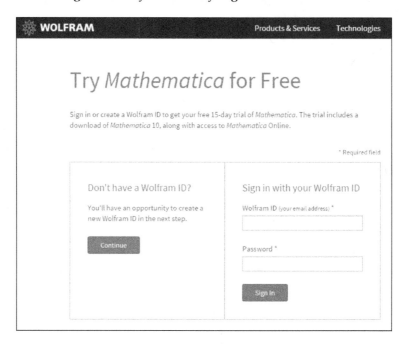

After registration, you will receive an e-mail containing the link to activate your Wolfram ID, as well as the link to access Mathematica Online. Activate your account and log in using `https://user.wolfram.com/portal/login.html`.

On this page, you can choose and download the Mathematica installer for your operating system.

[Copy the code in the **Activation Key** field. You will need it later to activate the product.]

Let's download the Windows version of the program (by clicking on the **Download for Windows** button) and go through the main stages of system installation on a computer. The size of the downloaded file will be slightly larger than 2 GB, so make sure you have enough free space on your computer, both for the installer and for the installation of Mathematica.

Once the Mathematica_10.2.0_WIN.zip file is downloaded to your computer, extract it into a separate directory and run setup.exe:

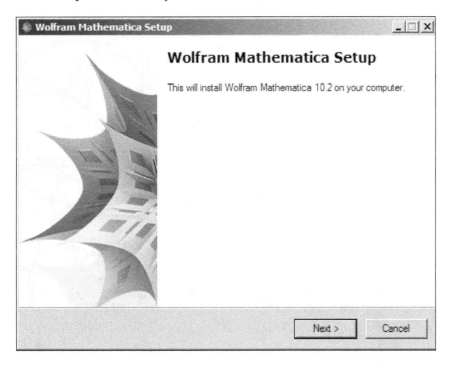

At this point, the installer will inform us that the installation process of Mathematica 10.2 is about to start. In order to proceed, click on **Next** in the preceding screenshot.

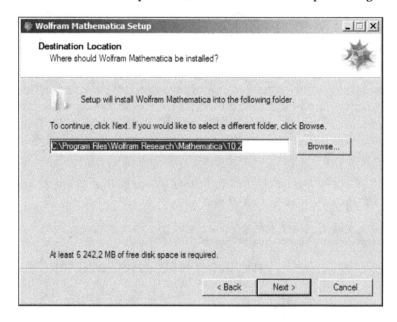

Note that the system installation requires 6.2 GB of free disk space. In the preceding screenshot, select the directory where you want to install Mathematica and click on **Next**.

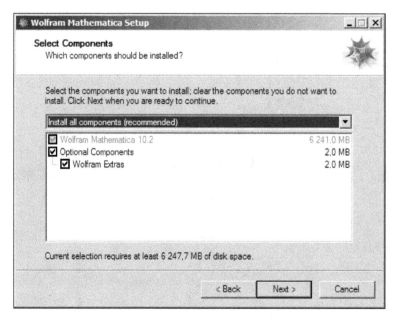

In the next dialog window, as shown in the preceding screenshot, you should choose the components of Mathematica you want to install. The Wolfram Extras package includes plugins for browsers that allow you to preview the Mathematica files embedded in web pages.

To continue the installation, click on the **Next** button.

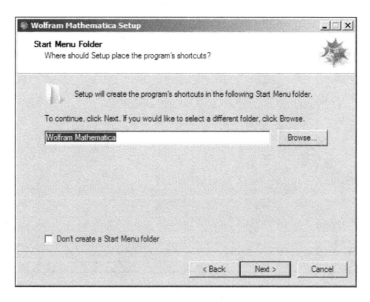

As shown in the previous screenshot, you need to select a folder in the Start Menu to place the program's shortcuts. Then click on **Next**.

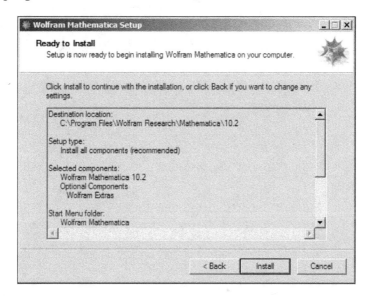

After that, we will see the summary of the previously selected options as shown in the preceding screenshot. Click on the **Install** button to start copying the files. After all files are copied, you will be notified that the installation is finished. You can check off the **Launch Wolfram Mathematica** checkbox to start learning the system immediately:

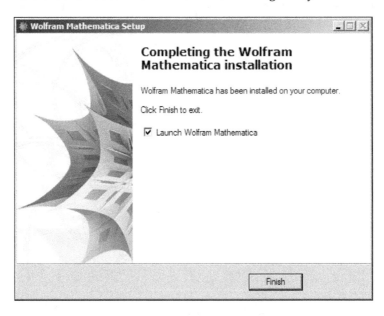

Once the system launches, you need to enter the activation key that we copied from the Wolfram website earlier:

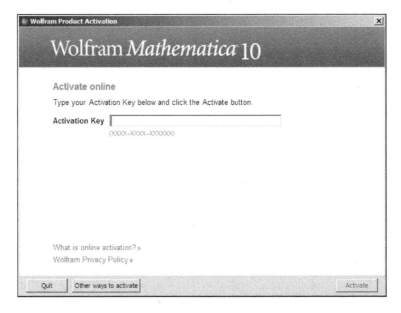

Now, click on the **Activate** button. You will be redirected to the following page:

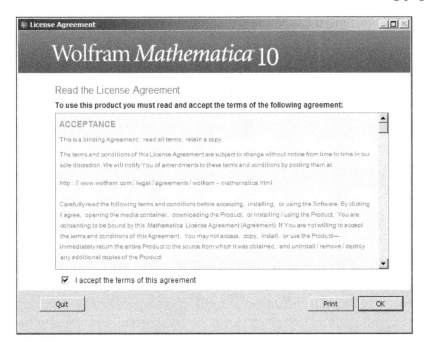

Read the license agreement and check the **I accept the terms of this agreement** box. Then click on **OK**.

Now, the Mathematica system is ready for the first use:

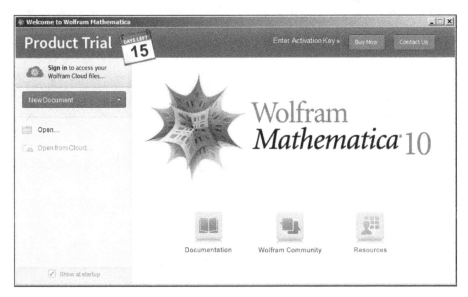

Setting up the system

To access the **Preferences** dialog, you need to select **Edit | Preferences...** in the application menu.

 On Mac OS X, the **Preferences** menu is located in the Mathematica application menu.

Let's review the most interesting settings of the system, which will be useful for our further work:

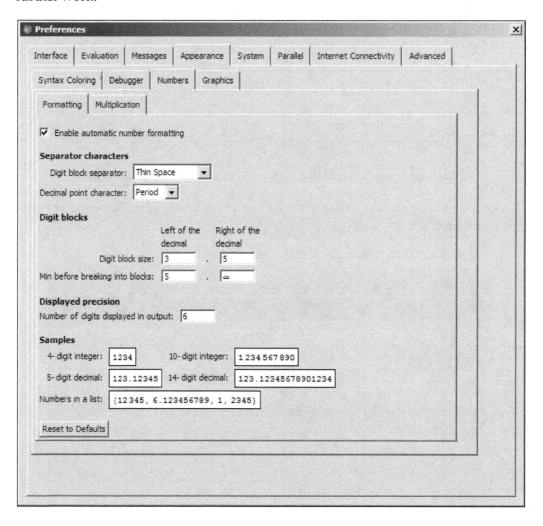

In the **Appearance | Number | Formatting** tab, you can select various numeric formatting options. Will there be a space separator between the digit blocks? What should be the decimal point character? How many digits will be displayed by default in the output?

In the next **Multiplication** tab, you can select the appearance of the multiplication:

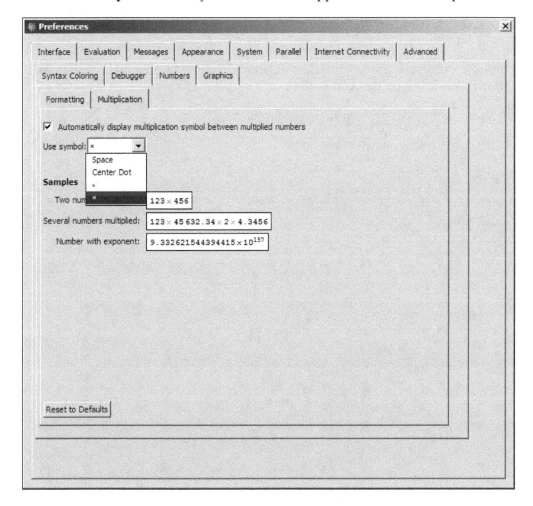

 In Mathematica, the space between two expressions means multiplying those expressions.

In this tab, you can choose whether the multiplication symbol between the multiplied numbers appears automatically, as well as choose the symbol type: space, center dot, *, or x.

The Mathematica front end and kernel

The Wolfram system has a modular structure and consists of two main parts. The one that directly performs all computations is called the **kernel**; the other part that interacts with the user is called the **front end**.

The kernel can be installed on the server, and it processes requests from multiple user interfaces. However, the single user interface can use several connected kernels for its computations.

In order to connect multiple kernels, navigate to **Evaluation | Kernel Configuration Options...** and click on the **Add** button.

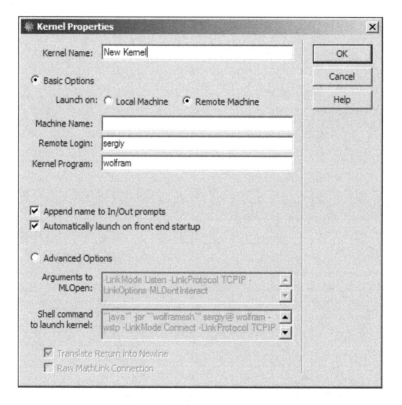

In the dialog window that opens, you can select whether the kernel will be located on a local computer or on a remote computer in the network. You can also enter the parameters in order to connect to this computer. There is a capability to be always aware of which kernel was used for computations.

The front end has the following interfaces:

- **Notebook**: It is the document in which the user enters all the necessary expressions and algorithms.

- **Text-based interface**: It is a command-line interface. It is used in text operating systems and not in the graphics operating systems.

- **Wolfram Symbolic Transfer Protocol (WSTP)**: It is the interface that enables interaction with other programs and modules.

Main features for writing expressions

At the beginning, it should be mentioned that Wolfram Mathematica has a very extensive reference system. You can access it by selecting in this menu: **Support | Wolfram Documentation**:

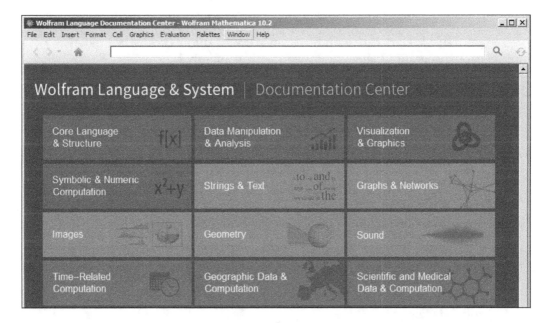

Let's get acquainted with some distinctions of Mathematica that will help us to understand the source code in the following chapters.

Let's create our first notebook. When Mathematica starts, you should choose **New Document** and then **Notebook**.

In order to compute any expression, press *Shift + Enter* after entering it. The input expression will be denoted by In and the output by Out:

```
In[1]:= 1 + 4
Out[1]= 5
```

 In this and other samples, all the formulas will start with In [Number]; you *shouldn't* type this because it is the number of the input that Mathematica calculates automatically.

All built-in functions *always* start with a capital letter, as shown here:

```
In[2]:= Cos [π / 4]

Out[2]=   1
        ─────
         √2
```

No variables *should* start with a number, since it will automatically be treated as a multiplication operation. They cannot end with a $ symbol.

If the variable ends with _, then it is a template expression and it can be substituted with anything, for example:

```
In[3]:= a[y_] := y y
        {a[2], a["apple"], a[π]}

Out[4]= {4, apple², π²}
```

In Mathematica, there are four types of brackets:

- **Square []**: These are used to describe the function parameters
- **Round ()**: These are used to group expressions and logical elements
- **Curly {}**: These are used to describe the elements of arrays, vectors, and data lists
- **Double Square [[ ]]**: These are used to allocate a specific item in a data list

In Mathematica, there are two types of assignments: **absolute** (=) and **delayed** (:=).

The difference is that in a delayed assignment, the right-hand side of the expression is always recomputed if the function is called. In an absolute assignment, the value that was in place during the assignment is stored.

```
In[8]:= x1 = TimeObject[Now]
        x2 := TimeObject[Now]

Out[8]=   ⊙ 22:01:15 GMT+3.

In[10]:= {x1, x2}
Out[10]=
          {  ⊙ 22:01:15 GMT+3.  ,   ⊙ 22:01:23 GMT+3.  }
```

In this example, we see that at the time of the second call of variables, x2 adopted the current time value. At the same time, x1 remained unchanged.

In order to use the result of the preceding expression, one can use the % symbol:

```
In[16]:= Log[Exp[π]]

Out[16]= π

In[17]:= Sin[% / 2]

Out[17]= 1
```

The basis of the Mathematica language is the functional form of all the expressions. In order to see a complete expression, the FullForm function should be used:

```
In[18]:= FullForm[(a + b) c]

Out[18]//FullForm=
              Times[Plus[a, b], c]
```

The variables that are not used should be cleared by the `Clear` function. Otherwise, computation errors will occur:

```
In[19]:= x = 5

Out[19]= 5

In[20]:= Range[x]

Out[20]= {1, 2, 3, 4, 5}

In[21]:= Solve[x^2 - 2 x + 1 == 0, x]

         Solve::ivar : 5 is not a valid variable. »

Out[21]= Solve[False, 5]

In[22]:= Clear[x]

In[23]:= Solve[x^2 - 2 x + 1 == 0, x]

Out[23]= {{x → 1}, {x → 1}}
```

Summary

In this chapter, we learned how to install the Wolfram Mathematica software and reviewed the basic components of the system: front end and kernel. We also gained a knowledge of the Mathematica language features.

In the next chapters, we will review additional features and techniques that facilitate work in Mathematica.

2
Broad Capabilities for Data Import

In order to process and analyze data, it is necessary to be able to import data into Mathematica. However, the data may not be always in a convenient format; they may contain errors, may be incomplete, or on the contrary—they may contain redundant information. Therefore, in order to learn how to import data, you will understand the following from this chapter:

- What types of data can be imported in Mathematica?
- What built-in functions are used for importing?
- How to check data for completeness
- How to clean the data and transform them into a convenient format for analysis

Permissible data format for import

Mathematica allows the importing of hundreds of data formats, but we only need some of them: those that are amenable to mathematical processing and analysis.

The main data formats can be divided into the following groups:

- **Tabular Text Formats**: This includes general text format data (*.dat), comma-separated (*.csv), or tab-separated (*.tsv)
- **Spreadsheet Formats**: This includes office programs data working with documents, such as Excel (*.xls and *.xlsx), Open Office (*.odc, and *.sxc), and even the first spreadsheet VisiCalc data (*.dif)
- **Data Interchange Formats**: This includes the JSON format

- **Database Formats**: This includes MS Access (*.mdb) and dBase (*.dbf)

- **Compression and Archive Formats**: This includes the files created by archivers, such as Windows ZIP (*.zip), Unix GZIP (*.gz), Unix TAR (*.tar), and BZIP2 (*.bz2)

- **XML/HTML Formats**: This includes extensions such as .xml, .xhtml, .html, and .rss

 Here, you may find the complete list of all the data formats that can be used for importing: http://reference.wolfram.com/language/guide/ListingOfAllFormats.html

Importing data in Mathematica

In order to import data into Mathematica, the Import function is used. Let's see how this feature is used in practice.

If you need to simply import the file, you can do it as follows:

```
Import["file_name"]
```

However, in practice, it is much more useful to use additional parameters of this function. Let's consider the following example.

Let's suppose we need to import the data on unemployment. We will take the information from the http://unstats.un.org/unsd/mbs/app/DataSearchTable.aspx page in the **Topic** list, select **EMPLOYMENT, UNEMPLOYMENT AND EARNINGS** and in the **Table** list select **11 Unemployment**. Then, select all the countries with the *Shift* button. After that download data by clicking on the **Get data** button and then clicking on the **Download in CSV** button, and save it into the C:\MathData\Unemployment.csv file.

In order to import data, let's select the Import function:

```
In[2]:= Import["C:\\MathData\\Unemployment.csv"]

Out[2]= {{country, country_code, series, seriescode,
         measure, period, D1, D2, D3, data, footnoteref},         ,
         {2046, Not in employment and currently available. For
         unemployment rate; must also be seeking employment., }}

large output   show less   show more   show all   set size limit...
```

Note that in Mathematica, \ is a special character. This is why in order to specify the path correctly, you should enter it as \\.

In the output, Mathematica has shown that it identified the file structure as a list consisting of two lists—the first one containing the actual data on unemployment in different countries and the second containing non-essential data—the explanations of the statistics.

If we want to import using the `Elements` parameter, then Mathematica interprets the incoming data, and in the output, separates them into the constituent elements.

```
In[3]:= Import["C:\\MathData\\Unemployment.csv", "Elements"]

Out[3]= {Data, Grid}
```

You can find out to which element a certain part of data belongs using the `Rules` parameter:

```
In[22]:= Import["C:\\MathData\\Unemployment.csv", "Rules"]

Out[22]= {Data → {{country, country_code, series, seriescode,
            measure, period, D1, D2, D3, data, footnoteref},        ,
          {2046, Not in employment and currently available. For unemployment
            rate; must also be seeking employment., }}, Grid →        }

         large output    show less    show more    show all    set size limit...
```

When importing a large amount of data, it is convenient to use the `Short` function to see an abridged version of the output. One of the function's parameters is the number of lines of the output data:

```
In[36]:= Short[Import["C:\\MathData\\Unemployment.csv"], 1]

Out[36]//Short= {{country, country_code, series, «5», D3, data, footnoteref}, «1878», {«1»}}
```

Since the first list of the imported data is column number, we can skip it using the `Rest` function. It will take all the data except the first one.

```
Short@(Rest[Import["C:\\MathData\\Unemployment.csv"]])
```

As we have a list of footnotes at the end of our imported data list, which we don't want to be analyzed, we will remove it by using the Cases function. This function only leaves the data in the list that match the pattern. In this case, the pattern will be any list that contains 11 elements (there are only 2 elements in the footnotes list). Let's look at the result of this function:

```
In[29]:=
    Short@Cases[Rest[Import["C:\\MathData\\Unemployment.csv"]],
        {x1_, x2_, x3_, x4_, x5_, x6_, x7_, x8_, x9_, x10_, x11_}]
Out[29]//Short= {{Azerbaijan, 31, «7», 237.2, /688}, «1855»}
```

 Note that Short@ denotes the output format and does not affect the data.

It is convenient to use the Cases function when you need only a part of the data; for example, you need to know about the levels of unemployment in Ukraine, as in the preceding example.

```
In[34]:=
    Cases[Rest[Import["C:\\MathData\\Unemployment.csv"]],
        {"Ukraine", _, _, _, _, _, _, _, _, _, _}]
Out[34]= {{Ukraine, 804, Labour force - Unemployment number, 809,
        (thousands),      2015JAN , Employment Office Records,
        Registered unemployment, number (thousands), 524.4, },
      {Ukraine, 804, Labour force - Unemployment number, 809,
        (thousands),      2014NOV , Employment Office Records,
        Registered unemployment, number (thousands), 450.6, },
      {Ukraine, 804, Labour force - Unemployment number, 809,
        (thousands),      2014DEC , Employment Office Records,
        Registered unemployment, number (thousands), 512.2, },
```

Suppose we need to analyze only the date and unemployment rate, and all other data are insignificant. Then, we need to leave only the columns with numbers 6 and 10. Let's remove them using the substitution expression /:

```
In[31]:= Short@Cases[Rest[Import["C:\\MathData\\Unemployment.csv"]],
        {x1_, x2_, x3_, x4_, x5_, x6_, x7_, x8_, x9_, x10_, x11_}] /.
        {{x1_, x2_, x3_, x4_, x5_, x6_, x7_, x8_, x9_, x10_, x11_} → {x6, x10}}
Out[31]//Short= {{     2015FEB , 237.2}, «1854», {     2014JUN , 6.8}}
```

In order to prepare this list for processing, it is necessary to transform the dates from the YYYYMMM format to the format of Mathematica with the help of the DateObject function, which will do the conversion from text to date. Besides, let's assign the result to the unempl variable so that you can return to it later.

```
In[22]:= unempl =
    Cases[Rest[Import["C:\\MathData\\Unemployment.csv"]],
       {x1_, x2_, x3_, x4_, x5_, x6_, x7_, x8_, x9_, x10_, x11_}] /.
      {{x1_, x2_, x3_, x4_, x5_, x6_, x7_, x8_, x9_, x10_, x11_} →
         {DateObject[x6], x10}}
```

Out[22]=
```
{{  Feb 2015 , 237.2},  {  Nov 2014 , 237.8},  {  May 2014 , 237.3},
 {  Aug 2014 , 238.2},  {  Feb 2015 , 4.9},  {  Aug 2014 , 4.9},
 {  Nov 2014 , 4.9},  {  May 2014 , 5.},  {         },  {  Oct 2014 , 6.4},
 {  Sep 2014 , 7.},  {  Aug 2014 , 7.},  {  Jul 2014 , 6.7},
 {  Apr 2014 , 7.1},  {  May 2014 , 7.1},  {  Jun 2014 , 6.8}}
```

Now we can work with this data. For example, we can construct an unemployment graph for the time period. Firstly, we need to group the data by date using the GroupBy function:

```
In[33]:=
    data = GroupBy[unempl, First → Last, Total]
```

Out[33]=
```
{  Feb 2015  → 70881.,     Nov 2014  → 67484.1,
   May 2014  → 75486.8,    Aug 2014  → 82357.7,    Apr 2015  → 37866.1,
   Mar 2015  → 54549.5,    Jan 2015  → 67644.,     Dec 2014  → 65633.2,
   Sep 2014  → 75077.,     Oct 2014  → 70865.8,    Jun 2014  → 75142.7,
   Jul 2014  → 79124.2,    Apr 2014  → 80831.5,    May 2015  → 13109.4 }
```

The first parameter of this function is a list, the second, that will be grouped together (in this case, the last part of the list—unemployment rate, will be grouped according to the first part of the list—by date), and the third parameter—a function that will be applied to the grouped data, that is, in this case its addition.

Now we just need to call `DateListPlot`, which will construct the required graph:

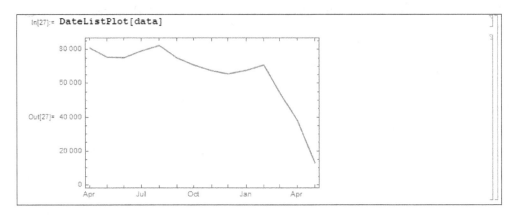

Additional cleaning functions and data conversion

Let's examine additional parameters of the functions that can help when importing data. Sometimes the data can be littered or have a wrong format—for example, instead of numeric, there may appear text data or it can be missing completely.

You can control the format of the incoming data with the help of templates that are used in the `Case` function. For example, if you need only numeric data to be imported, then the template will look like this: `x_?NumericQ`.

 All information about data types used in Mathematica can be found at `http://reference.wolfram.com/language/tutorial/PuttingConstraintsOnPatterns.html`.

Let's take a wider range of data for the analysis on unemployment in a country (country, date, and unemployment rate):

```
In[5]:= unempl =
    Cases[Rest[Import["C:\\MathData\\Unemployment.csv"]],
        {x1_, x2_, x3_, x4_, x5_, x6_, x7_, x8_, x9_, x10_?NumericQ, x11_}] /.
        {{x1_, x2_, x3_, x4_, x5_, x6_, x7_, x8_, x9_, x10_, x11_} →
            {x1, DateObject[x6], x10}}
```

Out[5]= {{Azerbaijan, 📅 Feb 2015 , 237.2}, {Azerbaijan, 📅 Nov 2014 , 237.8},

{Azerbaijan, 📅 May 2014 , 237.3}, ⟨⟨ 851 ⟩⟩,

{Venezuela (Bolivarian Republic of), 📅 May 2014 , 7.1},

{Venezuela (Bolivarian Republic of), 📅 Jun 2014 , 6.8}}

large output   show less   show more   show all   set size limit...

When you import data, it is important to check the format of the input data — for example, the number 2.3 can be identified as the string "2.3", and as a result, we are no longer able to work with strings. The input data format can be checked using the Head function:

```
In[7]:= Head /@ Rest[Import["C:\\MathData\\Unemployment.csv"]][[3]]

Out[7]= {String, Integer, String, Integer, String,
        String, String, String, Real, String}
```

In this case, the 10th parameter in the third string has a valid format, but to get reinsured and import all formats correctly, we'll use the NumberString function, which checks whether the string is a numerical data representation:

```
tst[x_] := If[StringQ[x], StringMatchQ[x, NumberString], NumericQ[x]]
```

We also use the If function that checks the first parameter on the validity and returns the second parameter in case it's true, or the third parameter — in case it's false.

Similarly, you can create a function that will clean up the extra characters and so on. For example, among the names of the countries, you can encounter extra brackets—Venezuela (Bolivarian Republic of) —let's remove them with a separate function:

```
In[37]:= cntr[y_] :=
        If[StringQ[y],
          StringReplace[y,
           StringCases[y, RegularExpression["\\s*\\((\\w*\\s*)*\\)"]] → ""], y]

In[38]:= cntr["Venezuela (Bolivarian Republic of)"]

Out[38]= Venezuela
```

In this function, the RegularExpression function is used, which is the parameter for the StringCases function. Thus, the result of the StringCases function will be all strings enclosed in brackets. Then, using the StringReplace function, all such strings are replaced by empty ones.

The details on the valid operators of the RegularExpression function can be found at http://reference.wolfram.com/language/ref/RegularExpression.html.

Checkpoint 2.1 – time for some practice!!!

Create a database of your expenses for the last one year. Try to plot a graph of your monthly, quarterly, and yearly expenses. Also, try to create RegularExpression that will help you to create a separate list of people to whom you have to pay money.

Once we have cleared the names of the countries from the unnecessary information, we can easily use them in the CountryData function in order to obtain additional information on the country, such as population, GDP, flag, and capital.

In general terms, with the functions used, the import will be represented like this:

```
In[61]:= unempl =
        Cases[Rest[Import["C:\\MathData\\Unemployment.csv"]],
          {x1_, x2_, x3_, x4_, x5_, x6_, x7_, x8_, x9_, x10_ /; tst[x10_], x11_}] /.
          {{x1_, x2_, x3_, x4_, x5_, x6_, x7_, x8_, x9_, x10_, x11_} →
            {CountryData[cntr[x1]], DateObject[x6], x10}}
```

Now, let's take unemployment information only for those countries that have data for January 2015. To do this, we'll write an additional function:

```
checkdate[x_] :=
If[DateObject[x] == DateObject["2015JAN"], True, False]
```

Since our data consists of two arrays — the first is data on the number of unemployed and the second is the percentage of unemployed, let's leave the data with the code 0809, which determines the number of unemployed:

```
checkrate[x_] := If[x == 0809, True, False]
```

As a result, we'll take only 10 such countries using Take:

```
In[83]:= unempl =
    Take[Cases[Rest[Import["C:\\MathData\\Unemployment.csv"]],
        {x1_, x2_, x3_, x4_ /; checkrate[x4], x5_, x6_ /; checkdate[x6],
        x7_, x8_, x9_, x10_ /; tst[x10], x11_}] /.
    {{x1_, x2_, x3_, x4_, x5_, x6_, x7_, x8_, x9_, x10_, x11_} →
        {CountryData[cntr[x1]], DateObject[x6],
        CountryData[cntr[x1], "Population"], x10}}, 10]
```

```
Out[83]= {{ Australia , 📅 Jan 2015 , 23 502 754 people , 838. },
    { Austria , 📅 Jan 2015 , 8 452 081 people , 406.2 },
    { Belgium , 📅 Jan 2015 , 10 841 093 people , 600.6 },
    { Belarus , 📅 Jan 2015 , 9 469 915 people , 30.7 },
    { Belarus , 📅 Jan 2015 , 9 469 915 people , 30.7 },
    { Canada , 📅 Jan 2015 , 35 309 555 people , 1312.6 },
    { Chile , 📅 Jan 2015 , 17 722 868 people , 519.6 },
    { Croatia , 📅 Jan 2015 , 4 369 961 people , 329.2 },
    { Cyprus , 📅 Jan 2015 , 1 153 409 people , 50. },
    { Czech Republic , 📅 Jan 2015 , 10 611 300 people , 556.2 }}
```

It should be noted that the data on the number of unemployed are measured in thousands of units and the population in units. That's why it's convenient to convert the data into numeric and remove the word people.

To demonstrate the Sort function, let's choose 10 countries with the highest number of unemployed people:

```
In[44]:=
        Short[Sort[unempl, #1[[3]] > #2[[3]] &], 10]

Out[44]//Short=  {{ Canada ,  📅 Jan 2015 ,  35 309 555 people , 1312.6},

                  { Australia ,  📅 Jan 2015 ,  23 502 754 people , 838.},

                  { Chile ,  📅 Jan 2015 ,  17 722 868 people , 519.6},

                  { Belgium ,  📅 Jan 2015 ,  10 841 093 people , 600.6},

                  { Czech Republic ,  📅 Jan 2015 ,  10 611 300 people , 556.2},

                  { Belarus ,  📅 Jan 2015 ,  9 469 915 people , 30.7},

                  { Belarus ,  📅 Jan 2015 ,  9 469 915 people , 30.7},

                  { Austria ,  📅 Jan 2015 ,  8 452 081 people , 406.2},

                  { Croatia ,  📅 Jan 2015 ,  4 369 961 people , 329.2},

                  { Cyprus ,  📅 Jan 2015 ,  1 153 409 people , 50.}}}
```

In this function, we used the templates of the #1 and #2 strings denoting the successive strings of data to be sorted. Since the elements of the sorted list are also lists, they can be accessed by index. To illustrate this, here's a simple example of sorting with all the on-screen elements that are being compared:

```
In[45]:= Sort[{4, 1, 5, 3, 6, 2}, (Print[{#1, #2}]; #1 > #2) &]
         {4, 1}
         {1, 5}
         {4, 5}
         {3, 6}
         {3, 2}
         {1, 6}
         {5, 6}
         {5, 3}
         {4, 3}
         {1, 3}
         {1, 2}
Out[45]= {6, 5, 4, 3, 2, 1}
```

Note that when using the patterns of the #1 and #2 strings, it is required that the function ends with the & symbol.

Examples examined in this section demonstrate how you can use Mathematica's information base in order to receive missing information during the analysis phase.

Importing strings

In particular, we should examine the ImportString function that allows you to import data from a certain format string. The following examples can help us trace function result changes depending on the type of input data.

```
In[84]:= ImportString["China,Chile,USA", "List"]
Out[84]= {China,Chile,USA}

In[86]:= ImportString["China\nChile\nUSA", "Table"]
Out[86]= {{China}, {Chile}, {USA}}

In[87]:= ImportString["China\nChile\nUSA", "Text"]
Out[87]= China
        Chile
        USA

In[93]:= ImportString["China<br/>Chile<br/>USA", "HTML"]
Out[93]= China
        Chile
        USA
```

 More information on valid formats for data import can be obtained by querying $ImportFormats in Mathematica.

Importing data from Mathematica's notebooks

In Mathematica, it is very convenient to transfer the formulas that you have used in other notebooks. This is done using the ImportNotebook function.

For example, let's create a new document using the CreateDocument function:

```
In[14]:= nb = CreateDocument[{TextCell["ImportDemo", "Title"],
         TextCell["First Input", "Section"],
         ExpressionCell[Defer[Sqrt[16]], "Input"],
         TextCell["Second Input", "Section"],
         ExpressionCell[Defer[Sqrt[-6]], "Input"]},
         WindowMargins → {{Automatic, 0}, {Automatic, 0}},
         WindowSize → {400, 400}];
      NotebookEvaluate[nb, InsertResults → True];
```

We have created a document in which there are only two cells to calculate — the root of 16 and the root of -6. This is a test example to demonstrate the function's capabilities. As a result, we get the following notebook:

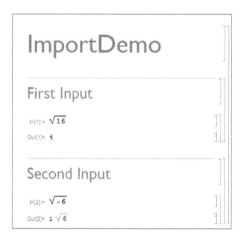

Let's consider import options with the help of the NotebookImport function. In order to import all the notebook's parameters, just call NotebookImport with the "_" parameter:

```
In[17]:= NotebookImport[nb, _]

Out[17]= {ImportDemo, First Input, HoldComplete[√16],
         4, Second Input, HoldComplete[√-6], i √6}
```

In order to convert the values to text, you can use the `_->"Text"` rule:

 Arrows and double square brackets are input with the help of the following combination of keys: *Esc - >Esc.*

```
In[18]:= NotebookImport[nb, _ → "Text"]

Out[18]= {ImportDemo, First Input, Sqrt[16], 4, Second Input, Sqrt[-6], I Sqrt[6]}
```

Let's use the `"Input"` parameter to get a list of cells with input data:

```
In[19]:= NotebookImport[nb, "Input"]

Out[19]= {HoldComplete[√16], HoldComplete[√-6]}
```

To import only the computed cells, you need to use the `"Input" -> "InputText"` rule:

```
In[20]:= NotebookImport[nb, "Input" → "InputText"]

Out[20]= {Sqrt[16], Sqrt[-6]}
```

To import both input and output cells, you can use a combination of the `"Input" | "Output"` parameters. The data will be grouped as desired, if you specify `"FlattenCellGroups" -> False`:

```
In[22]:= NotebookImport[nb, "Input" | "Output", "FlattenCellGroups" → False]

Out[22]= {{{HoldComplete[√16], 4}}, {{HoldComplete[√-6], i √6}}}}
```

To import the text containing the computation formulas, you should use the `"Input" | "Output" -> "InputText"` rule:

```
In[23]:= NotebookImport[nb, "Input" | "Output" → "InputText",
           "FlattenCellGroups" → False]|

Out[23]= {{{Sqrt[16], 4}}, {{Sqrt[-6], I Sqrt[6]}}}}
```

Controlling data completeness

In practice, it often happens that data is missing for some reason. If you take Mathematica's data about the countries of the world, not all of them will have updated information on population, GDP, and other parameters. In order to check the data for completeness, you can use the Missing function:

```
In[2]:= Count[Map[CountryData[#, "PopulationGrowth"] &, CountryData[All]],
          _Missing]

Out[2]= 2
```

This formula computes how many countries do not contain the information of the PopulationGrowth parameter. The Map function substitutes as an argument of the CountryData function values for all the countries that are stored in the database. The _Missing filter computes only those countries whose information is missing.

 There is also a short form of the Map - Map[f, {a, b, c}] function that can be written as f/@{a, b, c}. The result in both cases will be {f[a], f[b], f[c]}.

To prevent missing data that affects statistics, you can use the DeleteCases function:

```
In[3]:= dataset =
        DeleteCases[Map[CountryData[#, "PopulationGrowth"] &,
          CountryData[All]], _Missing]

Out[3]= {0.0330226 people/(person yr), 0.00339068 people/(person yr),
          0.0136518 people/(person yr), 0.01222 people/(person yr),
          0.00247996 people/(person yr), 0.0273764 people/(person yr),
          0.0214615 people/(person yr), 0.0127839 people/(person yr),
          0.00863338 people/(person yr), 0.00281444 people/(person yr),
          0.00272592 people/(person yr), 0.0128828 people/(person yr),
          0.00146496 people/(person yr), 0.0119105 people/(person yr),
          0.0112675 people/(person yr), 0.0130844 people/(person yr),
```

Thus, all the data used by Mathematica to find the `PopulationGrowth` parameter values and that fail to do this will be deleted from the initial sampling.

When working with time series, you can specify how the missing data will be processed. `TemporalData` represents a collection of paths composed of time-value pairs, {tij, vij}. And among its parameters, it contains `MissingDataMethod`, which can take the following values: `None` (when there's no action), `Automatic` (system automatically selects the best method), `Constant` (when there is a substitution of a constant value), `Interpolation` (with data interpolation), and other methods described by a user.

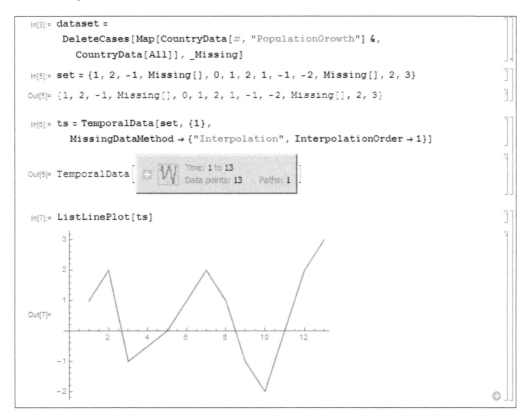

In this case, the linear interpolation1 was instead of the lost data,

```
InterpolationOrder->1
```

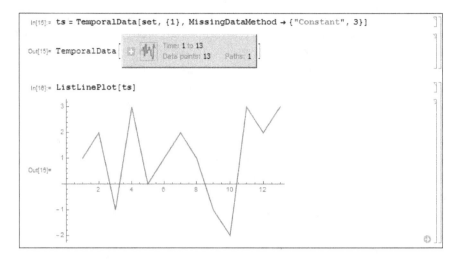

In this example, lost data was substituted by the constant 3.

When we work with the data received from external systems, their completeness control becomes a bigger problem. This happens because, in this case, instead of missing data, there can be a `NotAvailable` designation or the data is out of the set bounds.

Let's consider the following example. Suppose we have a set of data received from three groups: the first parameter is the number of the group, and the second and the third parameters are figures derived in this group:

```
In[18]:= d = {{1, 12.4, 0.51}, {3, 34.2, 5.92}, {1, 52.9, 1.38}, {2, 96., 1.74},
       {2, 41.1, 2.16}, {5, 19.1, 1.12}, {2, 81.1, 2.92}, {1, 43.1, 2.84},
       {1, 12.72, "-"}, {3, 34.4, 2.84}}

Out[18]= {{1, 12.4, 0.51}, {3, 34.2, 5.92}, {1, 52.9, 1.38},
       {2, 96., 1.74}, {2, 41.1, 2.16}, {5, 19.1, 1.12},
       {2, 81.1, 2.92}, {1, 43.1, 2.84}, {1, 12.72, -}, {3, 34.4, 2.84}}

In[20]:= Grid[d, Dividers → All]
```

1	12.4	0.51
3	34.2	5.92
1	52.9	1.38
2	96.	1.74
2	41.1	2.16
5	19.1	1.12
2	81.1	2.92
1	43.1	2.84
1	12.72	-
3	34.4	2.84

The `Grid` function allows us to provide the data in a convenient form. Now we can see that in the 6th string, we have data belonging to group 5, which is not available, and in the 9th string, there is no data on second observation.

Using the `MatchQ` function, which checks whether the expression corresponds to a given template, you can create a check to see whether the data is included in the 1st, 2nd, or 3rd group:

```
iswronggroup[x_] := Not[MatchQ[x[[1]], 1 | 2 | 3]]
```

This function will return the `False` value in an event in which the group number is not 1, 2, or 3, and it is not followed by numeric values.

Now we can remove incorrect data using the `DeleteCases` function:

```
In[25]:= cleaned = DeleteCases[d, _?iswronggroup]
Out[25]= {{1, 12.4, 0.51}, {3, 34.2, 5.92}, {1, 52.9, 1.38},
         {2, 96., 1.74}, {2, 41.1, 2.16}, {2, 81.1, 2.92},
         {1, 43.1, 2.84}, {1, 12.72, -}, {3, 34.4, 2.84}}
```

As we can see, group 5 is no longer in the input data.

However, there remains a problem with the missing information of sampling 8 new data elements. There might be several solutions; do not take this data into account (by adjusting the value of the `iswronggroup` function in order to make it react only to the `iswronggroup[x_] := Not[MatchQ[x, {1 | 2 | 3, _?NumberQ, _?NumberQ]]`) numeric data or to the approximate parameter data, for example by the average value of the group. Let's find the average value of the sampling second parameter in group 1:

```
In[26]:= gr1 = Select[cleaned, #[[1]] === 1 &]
Out[26]= {{1, 12.4, 0.51}, {1, 52.9, 1.38}, {1, 43.1, 2.84}, {1, 12.72, -}}

In[27]:= mgr1 = Mean[Select[gr1[[All, -1]], NumberQ]]
Out[27]= 1.57667
```

Using the `Select` function, we have chosen the data belonging to only group 1. With the help of the `Mean` function, we have computed the average value of the last column of data that has a numeric representation.

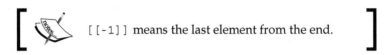 `[[-1]]` means the last element from the end.

```
In[28]:= cleaned[[All, 3]] = cleaned[[All, 3]] /. "-" → mgr1

Out[28]= {0.51, 5.92, 1.38, 1.74, 2.16, 2.92, 2.84, 1.57667, 2.84}

In[29]:= Grid[cleaned, Dividers → All]
```

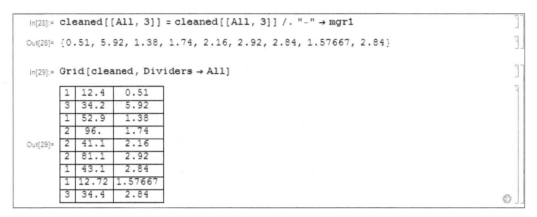

```
Out[29]=
```

1	12.4	0.51
3	34.2	5.92
1	52.9	1.38
2	96.	1.74
2	41.1	2.16
2	81.1	2.92
1	43.1	2.84
1	12.72	1.57667
3	34.4	2.84

Using the `/.` substitution statement, we have replaced the - value with the average value of group 1, and now the data has a valid form.

In practice, the solution of such a problem can look more elegant if you write a function that replaces the group values that have missing data with the selected method:

```
In[30]:= cleanByGroup[input_, col_, grpcol_, f_] :=
    Block[{columnvals, funvals, grouped, groups, rules},
      columnvals = input[[All, {grpcol, col}]];
      If[VectorQ[columnvals[[All, 2]], NumericQ], columnvals[[All, 2]],
        grouped = GatherBy[columnvals, First];
        groups = grouped[[All, 1, 1]];
        funvals = Table[f[Select[grouped[[i]][[All, 2]], NumericQ]],
          {i, Length[groups]}];
        rules = Table[Rule[{groups[[i]], _?(Not[NumericQ[#]] &)},
          {groups[[i]], funvals[[i]]}], {i, Length[groups]}];
        Replace[columnvals, rules, {1}][[All, 2]]]]

In[31]:= cleanByGroup[DeleteCases[d, _?iswronggroup], 3, 1, Mean]

Out[31]= {0.51, 5.92, 1.38, 1.74, 2.16, 2.92, 2.84, 1.57667, 2.84}
```

A user's procedures in the Mathematica environment are described in the Block construction, where the first parameter is the list of all variables used in the procedure and the second includes the instructions which go after ; .

Since the cleanByGroup function returns only the third column, let's attach it to the input data by replacing the previous one with the help of the Transponse function:

```
In[33]:= Grid[Transpose[Table[If[j == 1, ng[[All, j]], cleanByGroup[ng, j, 1, Mean]],
          {j, Length[ng[[1]]]}]]]

          1  12.4   0.51
          3  34.2   5.92
          1  52.9   1.38
          2  96.    1.74
Out[33]=  2  41.1   2.16
          2  81.1   2.92
          1  43.1   2.84
          1  12.72  1.57667
          3  34.4   2.84
```

Summary

In this chapter, we examined the basic functions that are used to import data into Mathematica. We have learned how to cast these data in a form convenient for analysis and to check them for errors and completeness. This knowledge will help us in the next chapter when we will need to retrieve data from external systems.

3
Creating an Interface for an External Program

If you are reading this book, then surely you are familiar with other data processing systems and have some store of knowledge consisting of your own programs, procedures, and computation methods. Having learned about the broad capabilities of Mathematica, you will certainly wish to transfer this experience into a new environment in order to make computations more accurate and effective. Many mathematical algorithms have been implemented in Mathematica, so you do not need to repeat the computations — it is enough to describe their logic of working with the input data.

In this chapter, you will learn the following:

- What Wolfram Symbolic Transfer Protocol is and how it can be used
- How to transfer algorithms and data structures previously written in languages such as C, C++, .NET, Java, or R into Mathematica
- How to use code written in Mathematica in an external program

Wolfram Symbolic Transfer Protocol

To make an external program's and the Mathematica kernel's instructions understandable to each other, we need a common protocol — and **WSTP** is such a protocol.

This protocol is used to include certain source code in the body of an external program, and then its internal functions and data structures are available in Mathematica.

In order to run an external program, you should use the `Install` function, where a path to the running program is used as a parameter. The `Uninstall` function is used in order to end a session with a remote program. For example, let's call one of the demo programs that come with the Mathematica system— the `bitops` and `bitAnd` functions, which return the conjunction of two integers:

In this example, we have used useful functions such as `SetDirectory` and `ResetDirectory` that allow us to set the folder to search for programs. Note the `$InstallationDirectory` variable—the folder in which Mathematica is installed, and the `$SystemID` variable—the system's version. We see that due to WSTP, we have obtained information about all the templates that are available for a call using `LinkPatterns [link]`, as well as the help `?bitAnd` - `bitAnd [x, y]` gives the bitwise conjunction of the two integers, x and y.

More globally, WSTP can be used for the following:

- Calling functions from external programs in the Mathematica environment
- Calling Mathematica's functions in external programs
- Data exchange between Mathematica and external programs
- Data exchange between parallel processes in Mathematica

With the help of WSTP, you can exchange the following data:

- An array of numbers
- A collection of geometric objects
- A command sequence
- A text flow
- Records in the database
- Cells from Mathematica's notebook

In order to make the external program understandable to WSTP, besides the internal settings specific to each programming language, it must also contain an appropriate WSTP template for each function in a file with the tm extension. For example, for the bitops program in the `C:\Program Files\Wolfram Research\Mathematica\10.1\SystemFiles\Links\MathLink\DeveloperKit\ Windows-x86-64\MathLinkExamples\bitops\bitops.tm` folder, the file looks like this:

```
int bitand P(( int, int));

:Begin:
:Function: bitand
:Pattern: bitAnd[x_Integer, y_Integer]
:Arguments: {x, y}
:ArgumentTypes: {Integer, Integer}
:ReturnType: Integer
:End:

:Evaluate: bitAnd::usage = "bitAnd[x, y] gives the bitwise
conjunction  of two integers x and y."
```

Let's take a closer look into this file:

`:Begin:`	Begin the template for a particular function
`:Function:`	The name of the function in the external program
`:Pattern:`	The pattern to be defined to call the function
`:Arguments:`	The arguments to the function
`:ArgumentTypes:`	The types of the arguments to the function
`:ReturnType:`	The type of the value returned by the function
`:End:`	End the template for a particular function
`:Evaluate:`	The Wolfram Language input to be evaluated when the function is installed

Then, this file is preprocessed using the `mprep` utility, and as an output, we receive a code written in C that contains everything we need for a call in the Mathematica environment.

Interface implementation with a program in C/C++

Let's consider how a C program should be prepared in order to make the WSTP protocol interaction available.

Firstly, as mentioned in the previous chapter, you need to prepare a special tm file that will describe all the data that will interact with Mathematica.

Secondly, the source code written in C should include the standard WSTP header file:

```
#include "wstp.h"
```

 This file is located in the `C:\Program Files\Wolfram Research\Mathematica\10.1\SystemFiles\Links\WSTP\DeveloperKit\Windows-x86-64\CompilerAdditions\mldev64\include` folder.

This file defines the following types:

WSLINK	A WSTP link object (analogous to LinkObject in the Wolfram Language)
WSMARK	A mark in a WSTP stream
WSENV	The WSTP library environment

All functions that return the `int` type—`return 0` if their execution is not successful.

Thirdly, the external program should be ready to receive instructions from the Mathematica environment; that's why the main function must have the following form:

```
int main(int argc, char *argv[]) {
  return WSMain(argc, argv);
}
```

After this, the received files are transferred as parameters to the mcc utility, and in the output, we get a compiled executable file that can accept requests from Mathematica.

 For more information on the parameters of the mcc utility, visit `http://reference.wolfram.com/language/ref/program/mcc.html`.

Besides being able to use external programs' data, Mathematica is also capable of generating and compiling C code. This is possible with the SymbolicC and CCompilerDriver packages. Let's consider an example of a simple C code that Mathematica can generate:

```
In[19]:= Needs["SymbolicC`"]

In[21]:= CBlock[{CDeclare["int", "x"], CAssign["x", COperator[Minus, {8, 3}]],
        CReturn["x"]}] // ToCCodeString

Out[21]= {
    int x;
    x = 8 - 3;
    return x;
}
```

In this example, we have enabled the SymbolicC package using the Needs function. Then, we have built the code saying that we want to create a function in C—the SBlock function, which will begin with the definition of the x variable (of the int type)—CDeclare["int", "x"]. The function body will consist of one assignment operation described using the CAssign function—the parameters for this are the x variable and the result of subtraction of the numbers 8 and 3 using COperator[Minus, {8, 3}]. The function returns the x: CReturn["x"] element. In order to have the result formatted in C-style, we have used the ToCCodeString function.

Similarly, you can create a C program using other functions of the SymbolicC package:

```
In[44]:= CProgram[CInclude["stdlib.h"], CInclude["constants.h"],
        CFunction["int", "sumPower", {{int, n}, {float, a}},
          CBlock[{CDeclare["int", "i"], CDeclare["float", "a"], CAssign[a, 0],
            CFor[CAssign[i, 0], COperator[Less, {i, n}], COperator[Increment, i],
              CAssign[AddTo, a, CStandardMathOperator[Power, {x, i}]]], CReturn["x"]}]] //
        ToCCodeString

Out[44]= #include "stdlib.h"

#include "constants.h"

int sumPower(int n, float a)
{
int i;
float a;
a = 0;
for( i = 0; i < n; i++)
{
a += pow(x, i);
}
return x;
}
```

 You can find more details about the functions of the package at
http://reference.wolfram.com/language/SymbolicC/
guide/SymbolicC.html

Calling Mathematica from C

You can also generate a complete library of C functions, which will call
Mathematica's functions, using the CCodeGenerate function from the
CCodeGenerator package:

```
In[45]:= Needs["CCodeGenerator`"]

In[46]:= c = Compile[{{x, _Real}}, Cos[x] + x^3 - 1/(1 + 2 x)]

Out[46]= CompiledFunction[          Argument count: 1
                                    Argument types: {_Real}  ]

In[47]:= file = CCodeGenerate[c, "myFunc"]

Out[47]= myFunc.c

In[48]:= FilePrint[file]

        =include "math.h"

        =include "WolframRTL.h"

        static WolframCompileLibrary_Functions funStructCompile;

        static mint I0_0;

        static mint I0_1;

        static mint I0_2;

        static mbool initialize = 1;

        =include "myFunc.h"

        DLLEXPORT int Initialize_myFunc(WolframLibraryData libData)
        {
        if( initialize)
        {
        funStructCompile = libData->compileLibraryFunctions;
        I0_2 = (mint) 2;
```

In this example, we have created and compiled a Cos[x] + x^3 - 1/(1 + 2x) function using the Compile function. To see the result of the code generation, we have called the FilePrint[f] function.

With the help of Mathematica, we can get a compiled file that is ready for launching. Let's download the MinGW-w64 compiler for 32-bit and 64-bit Windows at http://sourceforge.net/projects/mingw-w64/ and install it in the C:\mingw-w64\ folder. To use this, you need to enable the CCompilerDriver`GenericCCompiler` package and call the CreateExecutable function:

```
In[71]:= Needs["CCompilerDriver`GenericCCompiler`"]

In[72]:= hello = CreateExecutable[StringJoin[
          "#include <stdio.h>\n",
          "int main(){\n",
          "  printf(\"Hello world.\\n\");\n",
          "}\n"],
          "hello_world", "Compiler" → GenericCCompiler,
          "CompilerInstallation" →
            "C:\\mingw-w64\\x86_64-5.1.0-posix-seh-rt_v4-rev0\\mingw64\\bin",
          "CompilerName" → "x86_64-w64-mingw32-gcc.exe"];

In[73]:= Import["!\"" <> hello <> "\"", "Text"]

Out[73]= Hello world.
```

Please note that with the help of GenericCCompiler, we have indicated that we will use a nonstandard compiler and have specified the path to the executable file.

With these tools, Mathematica becomes a powerful environment to generate other programs.

In addition to the executable files in Mathematica, you can also connect dynamic libraries. Thus, all the functions of the library are available for Mathematica. You can exchange standard data types as well as whole expressions written in Mathematica.

```
In[1]:= FindLibrary["demo_linkobject"]

Out[1]= C:\Program Files\Wolfram
        Research\Mathematica\10.1\SystemFiles\Links\LibraryLink\LibraryResources\Windows
        -x86-64\demo_linkobject.dll

In[3]:= func = LibraryFunctionLoad["demo_LinkObject", "reverseString", LinkObject,
        LinkObject]

Out[3]= LibraryFunction[          Function name: reverseString
                                  Connection type: LinkObject      ]

In[4]:= func["Hello world"]

Out[4]= dlrow olleH
```

For example, let's consider the `demo_linkObject` library that comes with Mathematica. With the help of the `FindLibrary` function, we can see in which directory the library is located. The designation of this library is not only to show how you can call the `reverseString` function, but also to receive the result that is transferred to the Mathematica environment from this function. Using the `LibraryFunctionLoad` function, we have downloaded the `reverseString` function from the `demo_linkObject` library into Mathematica. The `LibraryFunctionLoad` function has the following arguments:

```
LibraryFunctionLoad[lib,fun,argtype,rettype]
```

lib	The library to be loaded
fun	The function name as specified in the library file
argtypes	The list of argument types
rettype	The return type

In this case, there is `LinkObject` at the input and the output of the function. It's an object that represents an active WSTP connection. The source code of the library is as follows:

```
#include "mathlink.h"
#include "WolframLibrary.h"
#include <stdlib.h>
#include <string.h>

DLLEXPORT mint WolframLibrary_getVersion( ) {
  return WolframLibraryVersion;
}

static MTensor tensor;

DLLEXPORT int WolframLibrary_initialize( WolframLibraryData
libData) {
  return 0;
}

static char* reverseStringImpl( const char* inStr)
{
  int i, len;
  char* outStr;

  len = strlen( inStr);
```

```
  outStr = (char*)malloc( len+1);

  outStr[len] = '\0';
  for ( i = 0; i < len; i++) {
    outStr[i] = inStr[len-i-1];
  }
  return outStr;
}

DLLEXPORT int reverseString( WolframLibraryData libData, MLINK
mlp)
{
  int res = LIBRARY_FUNCTION_ERROR;
  int i1, i2, sum;
  long len;
  const char *inStr = NULL;
  char* outStr = NULL;

  if ( !MLTestHead( mlp, "List", &len))
    goto retPt;
  if ( len != 1)
    goto retPt;

  if(! MLGetString(mlp, &inStr))
    goto retPt;

  if ( ! MLNewPacket(mlp) )
    goto retPt;

  outStr = reverseStringImpl(inStr);

  if (!MLPutString( mlp,outStr))
    goto retPt;
  res = LIBRARY_NO_ERROR;
retPt:
  if ( inStr != NULL)
    MLReleaseString(mlp, inStr);
  if ( outStr != NULL)
    free(outStr);
  return res;
}
```

Note the `mathlink.h` and `WolframLibrary.h` connected headers that describe the data types used to call the library from the Mathematica environment. As you can see, among the `reverseString` function parameters, there is `MLINK` — the argument that uses the WSTP API to read the arguments that come in a list. After it has generated the result, this is written onto the link.

 A more detailed guide on how to compile and run Wolfram Symbolic Transfer Protocol (WSTP) programs written in the C language on computers running the Microsoft Windows operating system can be found at `http://reference.wolfram.com/language/tutorial/WSTPDeveloperGuide-Windows.html`.

Interacting with .NET programs

One of the distinctive features of using .NET in Mathematica is that in order to use all the capabilities of .NET, we don't need to write code at all. You can write a program right within the Mathematica environment. In this section, you will learn how to download .NET assemblies and types directly to Mathematica, as well as how to create objects from these types, call methods, properties, and so on.

In order to enable Mathematica's functions to interact with .NET, you need to download the `NETLink` package:

```
Needs["NETLink`"]
```

To run the .NET environment, it's necessary to run `InstallNET[]`. In case of an active development and a change of .NET classes to restart the environment, it's convenient to run `ReinstallNET[]`.

Using `LoadNETAssembly`, we will load a .NET assembly into the Mathematica environment:

```
LoadNETAssembly["System.Web"]
```

The .NET types are loaded with the help of the `LoadNETType` function:

```
LoadNETType["System.Windows.Forms.Form"]
```

To demonstrate the principles of interaction between Mathematica and .NET, let's create a modal window. Using the NETNew function, we will create a System.Windows.Forms.Form class object:

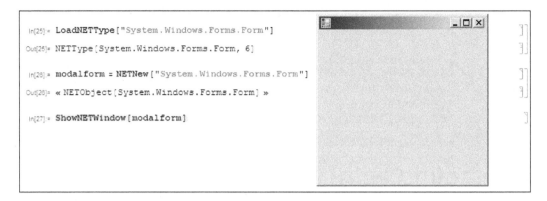

```
In[25]:= LoadNETType["System.Windows.Forms.Form"]

Out[25]= NETType[System.Windows.Forms.Form, 6]

In[26]:= modalform = NETNew["System.Windows.Forms.Form"]

Out[26]= « NETObject[System.Windows.Forms.Form] »

In[27]:= ShowNETWindow[modalform]
```

To show the modal window created, we have used the ShowNETWindow function. One of the advantages of using Mathematica for programming interfaces is that you can experiment with the program simultaneously when it is running. For example, you can easily change the window background color in real time simply by changing the BackColor property:

```
In[33]:= LoadNETType["System.Drawing.Color"];
         modalform@BackColor = Color`Blue;
```

To make it more interactive, let's add the onClick event processing by means of Mathematica. To do this, let's call the AddEventHandler function and associate the Click event with the OnClick procedure. Thus, each time a user clicks the window, it will change the background color:

```
In[35]:= AddEventHandler[modalform@Click, onClick]

Out[35]= « NETObject[System.EventHandler] »

In[36]:= onClick[args___] :=
         modalform@BackColor = Color`FromArgb[Random[Integer, {0, 255}],
           Random[Integer, {0, 255}], Random[Integer, {0, 255}]]
```

In order to make the window modal, just call DoNETModal[modalform]. Thus, as long as the window is not closed, this function will not return a value.

To put everything in one program module, it's convenient to use the `NETBlock`
function. Let's extend the window's functionality by adding a menu and a color
selection option:

```
In[52]:= SimpleModal[] :=
    NETBlock[Module[{modalform, onClick},
      modalform = NETNew["System.Windows.Forms.Form"];
      modalform@Width = 200;
      modalform@Height = 200;
      modalform@Text = "This is .NET";
      selectcolor = NETNew["System.Windows.Forms.MenuItem"];
    selectcolor@Text = "Select Background Color";
      menu = NETNew["System.Windows.Forms.MenuItem"];
    menu@Text = "Menu";
      menu@MenuItems@AddRange[{{selectcolor}}];
      mainmenu = NETNew["System.Windows.Forms.MainMenu", {menu}];
      modalform@Menu = mainmenu;
      AddEventHandler[selectcolor@Click, selecttextcolorclick];
      selecttextcolorclick[_, _] :=
        (ColorDialog = NETNew["System.Windows.Forms.ColorDialog"];
         ColorDialog@Color = modalform@BackColor;
         LoadNETType["System.Windows.Forms.DialogResult"];
         If[ColorDialog@ShowDialog[] === DialogResult`OK,
           modalform@BackColor = ColorDialog@Color]);
    DoNETModal[modalform]]]
In[53]:= SimpleModal[]
```

Let's consider Mathematica's capabilities in .NET applications in the example of the
`SimpleLink` program, which is included in the Mathematica distribution kit:

```
using System;
using Wolfram.NETLink;
public class SimpleLink {

  public static void Main(String[] args) {

    // This launches the Mathematica kernel:
    IKernelLink ml = MathLinkFactory.CreateKernelLink();

    // Discard the initial InputNamePacket the kernel will send
    when launched.
    ml.WaitAndDiscardAnswer();

    // Now compute 2+2 in several different ways.

    // The easiest way. Send the computation as a string and get
    the result in a single call:
```

```
        string result = ml.EvaluateToOutputForm("2+2", 0);
        Console.WriteLine("2 + 2 = " + result);

        // Use Evaluate() instead of EvaluateToXXX() if you want to
        read the result as a native type
        // instead of a string.
        ml.Evaluate("2+2");
        ml.WaitForAnswer();
        int intResult = ml.GetInteger();
        Console.WriteLine("2 + 2 = " + intResult);

        // You can also get down to the metal by using methods from
        IMathLink:
          ml.PutFunction("EvaluatePacket", 1);
          ml.PutFunction("Plus", 2);
          ml.Put(2);
          ml.Put(2);
          ml.EndPacket();
          ml.WaitForAnswer();
          intResult = ml.GetInteger();
          Console.WriteLine("2 + 2 = " + intResult);

        // Always Close link when done:
        ml.Close();

        // Wait for user to close window.
        Console.WriteLine("Press Return to exit...");
        Console.Read();
    }
  }
```

To get started, you should enable the `Wolfram.NETLink` library that includes the description of all the necessary classes. Then, the `CreateKernelLink()` Mathematica kernel is launched; as this process may take some time, the `WaitAndDiscardAnswer()` function is used. Next, there is a transfer of formulas for result computation and processing, which will be returned by the Mathematica kernel. After completion of work, it is necessary to close the connection with the Mathematica kernel using the `Close()` function.

You can compile this program code into an executable file using the command-line C# compiler, which is included for free with the .NET Framework SDK. It is necessary to make sure that the directory in which it is located is available for the PATH operating system variable. Besides, don't forget to copy the `Wolfram.NETLink.dll` library to the directory where the `SimpleLink.cs` source code is located. The compilation is carried out with the following parameters:

```
csc /target:exe /reference:Wolfram.NETLink.dll SimpleLink.cs
```

More details about interacting with .NET programs can be found at http://reference.wolfram.com/language/NETLink/ tutorial/Overview.html.

Interacting with Java

Having mastered the principles of Mathematica's interaction with external programs, it will be easy to understand the principles of writing Java programs in Mathematica.

First, you need to enable the JLink package. Here's an example Java program that will produce an expression computation using Mathematica's capabilities:

```
In[81]:= Needs["JLink`"]

      createWindow[] := Module[{frame, slider, listener}, InstallJava[];
        inText = JavaNew["java.awt.TextArea", "Expand[(x+1)^a]", 8, 40];
        outText = JavaNew["java.awt.TextArea", 8, 40];
        frame = JavaNew["com.wolfram.jlink.MathFrame", "RealTimeAlgebra"];
        JavaBlock[frame@setLayout[JavaNew["java.awt.BorderLayout"]];
         slider = JavaNew["java.awt.Scrollbar", Scrollbar`HORIZONTAL, 0, 1, 0, 20];
         frame@add[slider, ReturnAsJavaObject[BorderLayout`NORTH]];
         frame@add[outText, ReturnAsJavaObject[BorderLayout`CENTER]];
         frame@add[inText, ReturnAsJavaObject[BorderLayout`SOUTH]];
         frame@pack[];
         outText@setFont[JavaNew["java.awt.Font", "Courier", Font`PLAIN, 12]];
         listener = JavaNew["com.wolfram.jlink.MathAdjustmentListener"];
         listener@setHandler["adjustmentValueChanged", "sliderFunc"];
         slider@addAdjustmentListener[listener];
         frame@setLocation[200, 200];
         JavaShow[frame];];
        frame]

      sliderFunc[evt_, type_, scrollPos_] :=
       outText@setText[Block[{a = scrollPos}, ToString[ToExpression[inText@getText[]]]]]

      RealTimeAlgebraModal[] := JavaBlock[Module[{frm}, frm = createWindow[];
        frm@setModal[];
        DoModal[];]]

In[89]:= RealTimeAlgebraModal[]
```

Similar to .NET, the following functions are used: JavaBlock for procedure construction in Java language and JavaNew to define a new object copy. The Moving slider event is declared with the help of the setHandler method.

After running the `RealTimeAlgebraModal[]` command, you will see the following window where you can change the parameter value and a power series in the expansion by moving the slider:

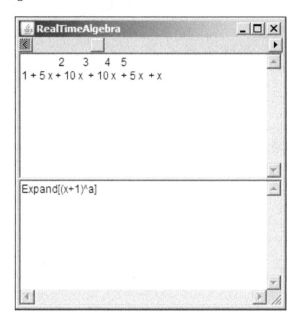

Let's look at a simple example of how to use the computing capabilities of Mathematica in a program written in Java:

```
import com.wolfram.jlink.*;

public class SampleProgram {

  public static void main(String[] argv) {

    KernelLink ml = null;

    try {
      ml = MathLinkFactory.createKernelLink(argv);
    } catch (MathLinkException e) {
      System.out.println("Fatal error opening link: " +
      e.getMessage());
      return;
    }

    try {
      // Get rid of the initial InputNamePacket the kernel will
      send
```

```
        // when it is launched.
        ml.discardAnswer();

        // This demonstrates a simple way to send a computation.
        Very useful if it is convenient to
        // create the input as a Java string. Note that the result
        is thrown away by discardAnswer().
        ml.evaluate("<<MyPackage.m");
        ml.discardAnswer();

        // This demonstrates how to read a result.
        ml.evaluate("2+2");
        ml.waitForAnswer();
        int result = ml.getInteger();
        System.out.println("2 + 2 = " + result);

        // Here's how to send the same input, but not as a string:
        ml.putFunction("EvaluatePacket", 1);
        ml.putFunction("Plus", 2);
        ml.put(3);
        ml.put(3);
        ml.endPacket();
        ml.waitForAnswer();
        result = ml.getInteger();
        System.out.println("3 + 3 = " + result);

        // If you want the result back as a string, use
        evaluateToInputForm or
        // evaluateToOutputForm. The second arg for either is the
        requested page
        // width for formatting the string. Pass 0 for
        PageWidth->Infinity.
        // These methods get the result in one step--no need to call
        waitForAnswer.
        String strResult = ml.evaluateToOutputForm("4+4", 0);
        System.out.println("4 + 4 = " + strResult);

    } catch (MathLinkException e) {
        System.out.println("MathLinkException occurred: " +
        e.getMessage());
    } finally {
        ml.close();
    }
  }
}
```

In order to use the classes that allow you to establish a connection with Mathematica, utilize `com.wolfram.jlink.*`. The connection is done using the `createKernelLink` function, and it is closed with the help of the `close` function. The computation of an expression is done with the help of the `evaluate` function, as well as the results analysis.

If you wish to compile this program for Windows, use a line like this:

```
javac -classpath ".;\path\to\JLink.jar" SampleProgram.java
```

 You can find more detailed instructions on interacting with Java at http://reference.wolfram.com/language/JLink/tutorial/Overview.html.

Interacting with R

If you are familiar with the R programming language and the environment for statistical studies, then you can easily transfer the accumulated material for use in Mathematica. To do this, you need to enable the `RLink` package and download the **paclet** for R using `RLinkResourcesInstall`:

```
In[91]:= Needs["RLink`"]

In[92]:= RLinkResourcesInstall[]

Out[92]= {Paclet[RLinkRuntime, 9.0.0.0, <>]}
```

When launching the `RLinkResourcesInstall` function, you will be asked to confirm the installation, and after this, the process of downloading the necessary files will start.

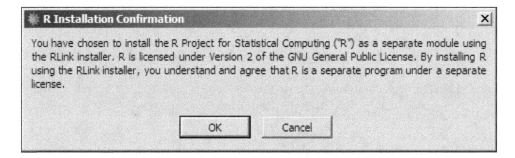

To start a connection with R, you need to call the `InstallR` function. For example, let's set the value of the `myR` variable using the `RSet` function and then read it using the `REvaluate` function:

```
In[95]:= InstallR[]

In[106]:= RSet["myR", "15"]

Out[106]= 15

In[107]:= REvaluate["myR"]

Out[107]= {15}
```

With the help of the `REvaluate` function, you can also declare functions written in R and execute them in Mathematica. For example, let's declare a function that returns the cube of given numbers:

```
In[111]:= thrd = REvaluate["f <- function(var) var^3"]

Out[111]= RFunction[closure, RCode[function (var)
         var^3], 2, RAttributes[]]

In[112]:= thrd[{2, 4, 6}]

Out[112]= {8., 64., 216.}
```

To avoid the repeated calling of the `REvaluate` function every time the `thrd` function is called, you can use `RFunction`:

```
In[113]:= thrdnew = RFunction["function(var) var^3"]

Out[113]= RFunction[closure, RCode[function(var) var^3], Automatic, RAttributes[]]

In[114]:= thrdnew[{2, 4, 6}]

Out[114]= {8., 64., 216.}

In[115]:= RSet["thrdnew", thrdnew]

Out[115]= RFunction[closure, RCode[function(var) var^3], Automatic, RAttributes[]]

In[116]:= REvaluate["thrdnew(c(2,4,5))"]

Out[116]= {8., 64., 125.}
```

In this example, we will also transfer the `thrdnew` function to R, where we have computed the result using the same data set.

So, you can see that with the help of R, it's easy to transfer data and perform computations.

In order to finish the connection session with R, you need to call the `UninstallR` function.

 You can get more detailed instructions about interacting with R at `http://reference.wolfram.com/language/RLink/tutorial/UsingRLink.html`.

Summary

In this chapter, we acquired the basic skills to transfer accumulated data-processing tools to Mathematica, as well as to use Mathematica's capabilities in computing expressions in other systems.

In the next chapter, we will learn some techniques to cluster, classify, and identify the various types of data.

4

Analyzing Data with the Help of Mathematica

Now we've come to one of the most interesting parts of this multifaceted Mathematica system. A bunch of already implemented and working algorithms are waiting for you to be used to build fast and compact solutions for set tasks.

In this chapter, you will learn the following:

- Clustering information and classifying data
- Recognizing an object in a picture and reckoning it in a particular category
- Identifying people's faces in a photo
- Recognizing text information and determining the language of the text
- Reading the different types of barcodes

Data clustering

Clusters are data groups of elements that are very close or similar. For example, a group of people can be divided into clusters according to age, height, sex, social status, and so on. Clustering helps to better understand input information because if we know the properties of one element of the cluster, it is likely that the other elements may also have these properties. The process of finding a cluster can go on without a teacher (unsupervised learning technique) and can be based on two functions: the **distance** function that indicates the distance between the elements of a cluster — the closer the elements are to each other, the greater is the probability that they are in the same cluster, and the **dissimilarity** function, the result of which is the degree of dissimilarity between the elements.

To cluster data, we'll use the `FindClusters` function. First, let's consider its application in simple examples:

```
In[4]:= list = {1, -2, 11, 17, 6, 3, 15, 21}
Out[4]= {1, -2, 11, 17, 6, 3, 15, 21}

In[5]:= FindClusters[list]
Out[5]= {{1, -2, 6, 3}, {11, 17, 15, 21}}

In[7]:= FindClusters[list, 3]
Out[7]= {{1, -2, 3}, {11, 6}, {17, 15, 21}}
```

By default, the `FindClusters` function finds clusters on the basis of the shortest distance. Since we have not specified how many clusters we would like to find, the result of running `FindClusters[list]` is two clusters: {1, -2, 6, 3} and {11, 17, 15, 21}. In the second case, we have clearly indicated that the data list contains three clusters, and based on this, the function has found them.

Let's generate three sets of random points distributed by a normal distribution and see how the `FindClusters` function will divide them:

```
randomND[s_] := RandomReal[NormalDistribution[0, s]]

randomData[n_Integer, p_, sigma_] :=
   Table[p + {randomND[sigma], randomND[sigma]}, {n}];
data = BlockRandom[SeedRandom[1100];
   Join[randomData[50, {3.5, 3}, .3], randomData[50, {2, 1.5}, .4],
     randomData[50, {1, 2.1}, .1], randomData[50, {3.75, 1.75}, 0.2]]];

clst = FindClusters[data];

ListPlot[clst]
```

As you can see, we have got four quite clear data sets.

To fine-tune the recognition of clusters, you can use advanced settings. One of them is the function of distance, which should satisfy the following conditions:

- $f(e_i, e_i)=0$
- $f(e_i, e_j)≥0$
- $f(e_i, e_j)= f(e_j, e_i)$

Mathematica supports the following distance functions:

- `EuclideanDistance[u,v]`: The Euclidean norm, $\sqrt{\sum(u-v)^2}$
- `SquaredEuclideanDistance[u,v]`: The squared Euclidean norm, $\sum(u-v)^2$
- `ManhattanDistance[u,v]`: The Manhattan distance, $\sum|u-v|$
- `ChessboardDistance[u,v]` : The chessboard or Chebyshev distance, $\max(|u-v|)$
- `CanberraDistance[u,v]` : The Canberra distance, $\sum|u-v|/(|u|+|v|)$
- `CosineDistance[u,v]` : The cosine distance, $1-u.v/(\|u\|\|v\|)$
- `CorrelationDistance[u,v]` : The correlation distance, `1-(u-Mean[u]).(v-Mean[v])/(Abs[u-Mean[u]]Abs[v-Mean[v]])`
- `BrayCurtisDistance[u,v]`: The Bray–Curtis distance, $\sum|u-v|/\sum|u+v|$

In order to use a specific function, simply specify it in the parameters as `FindClusters[data, DistanceFunction -> ManhattanDistance]`.

You can also specify the method of finding clusters (`Agglomerate` or `Optimize`), conduct a test to determine the best number of clusters (the `SignificanceTest` parameter), as well as specify the clustering linkage to be used (the `Linkage` parameter):

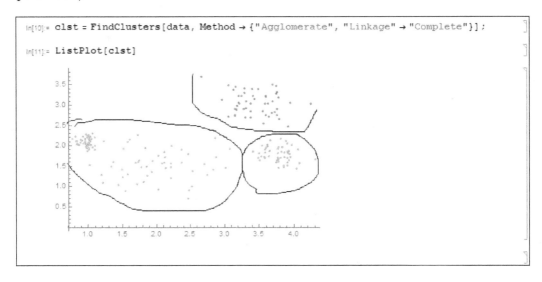

In this case, we have used the agglomerative method with the largest intercluster dissimilarity and got three clusters.

Similarly, you can partition into cluster arrays with more dimensions and various types of data: text, Boolean, and numeric.

Data classification

If clustering relates to learning without a teacher, then **classification**, on the contrary, is knowing to what groups a part of the known data belongs, and we want to determine the probability with which the unknown new element might belong to one group or another.

For example, using the `Classify` function, let's try to explain which are even numbers and which are odd numbers:

```
In[126]:= numbers = {1 → "odd", 2 → "even", 3 → "odd", 4 → "even",
            7 → "odd", 8 → "even", 9 → "odd"}

Out[126]= {1 → odd, 2 → even, 3 → odd, 4 → even, 7 → odd, 8 → even, 9 → odd}

In[127]:= c = Classify[numbers]

Out[127]= ClassifierFunction
```

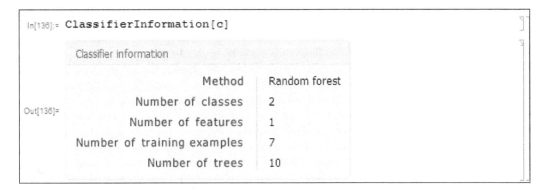

```
In[130]:= {c[5], c[0], c[10]}

Out[130]= {odd, even, even}
```

We have set several even and several odd numbers, then we have classified them with default parameters using the Classify function, and finally we can see how the missing elements, 5, 0 and 10, will be classified. As you can see from the result, they were successfully and correctly defined:

```
In[134]:= c[12, "Probabilities"]

Out[134]= <|even → 0.584659, odd → 0.415341|>
```

With the help of the Probabilities parameter, you can determine how likely it is for an element to belong to a particular class.

The ClassifierInformation function provides information about the sample data based on which the classification took place:

```
In[136]:= ClassifierInformation[c]
```

Classifier information

Method	Random forest
Number of classes	2
Number of features	1
Number of training examples	7
Number of trees	10

Out[136]=

In Mathematica, there are also built-in classes from different areas of knowledge:

CountryFlag	Which country a flag image is for
FacebookTopic	Which topic a Facebook post is about
Language	Which natural language the text is in
NameGender	Which gender a first name is
Profanity	Whether the text contains profanity
Sentiment	The sentiment of a social media post
Spam	Whether an email is spam

In order to use one of these classes, you should specify its name as the first parameter of the Classify function:

```
In[137]:= Classify["FacebookTopic",
              "It's official: Twin Peaks is coming back and will air
                  on Showtime Networks in 2016! "]

Out[137]= Television

In[152]:= Classify["NameGender", "Kyle"]

Out[152]= Male

In[165]:= Classify["Spam",
              "I bring forth a
                  business proposal in the tune of Ten Million Five Hundred
                  Thousand United State Dollars Only."]

Out[165]= True
```

Let's consider how we can use the data obtained with the help of the Classify function. For example, let's take a set of data that contains a list of Titanic passengers with their age, sex, ticket class, and survival and classify this using the logistic regression method:

```
In[166]:= titanic = ExampleData[{"MachineLearning", "Titanic"}, "Data"];

In[168]:= c = Classify[titanic, Method → "LogisticRegression"]

Out[168]= ClassifierFunction[       Method: LogisticRegression
                                      Number of classes: 2      ]
```

Now we'll construct the survival probability graph depending on the passenger's sex and the class in which they travelled:

```
In[169]:= prob[class_, age_, sex_] :=
            c[{class, age, sex}, {"Probability", "survived"}];

In[173]:= Plot[{prob["1st", x, "female"], prob["3rd", x, "female"],
            prob["1st", x, "male"], prob["3rd", x, "male"]}, {x, 0, 100},
          PlotLegends → {"female, 1st class", "female, 3rd class",
            "male, 1st class", "male, 3rd class"}, Frame → True,
          FrameLabel → {"Age (years)", "Survival probability"},
          PlotStyle → {Dashing[0.02], Dashing[0.04], Dashing[0.07],
            Dashing[1]}]
```

Out[173]=

Based on the graphical representation, we see that women from the first class had a very high probability of survival and men from the third class had very low probability. Besides, women from the third class were offered places by men from the first class.

The `Classify` function is also useful to determine an author's handwriting. For example, let's take three classic works by Shakespeare, Oscar Wilde, and Victor Hugo:

```
In[174]:= Othello =
        Import["http://www.gutenberg.org/cache/epub/2267/pg2267.txt"];
      Hamlet =
        Import["http://www.gutenberg.org/cache/epub/2265/pg2265.txt"];
      Macbeth =
        Import["http://www.gutenberg.org/cache/epub/2264/pg2264.txt"];

In[178]:= TheImportanceOfBeingEarnest =
        Import["http://www.gutenberg.org/cache/epub/844/pg844.txt"];
      ThePictureofDorianGray =
        Import["http://www.gutenberg.org/cache/epub/174/pg174.txt"];
      AnIdealHusband =
        Import["http://www.gutenberg.org/files/885/885-0.txt"];

In[181]:= LesMiserables =
        Import["http://www.gutenberg.org/cache/epub/135/pg135.txt"];
      NotreDamedeParis =
        Import["http://www.gutenberg.org/cache/epub/2610/pg2610.txt"];
      TheManWhoLaughs =
        Import["http://www.gutenberg.org/cache/epub/12587/pg12587.txt"];
```

We'll use these works to educate the `Classify` function and then present the third work and see whether it manages to guess the author:

```
In[184]:= a = Classify[<|"William Shakespeare" → {Othello, Hamlet},
          "Oscar Wilde" → {TheImportanceOfBeingEarnest,
            ThePictureofDorianGray},
          "Victor Hugo" → {LesMiserables, NotreDamedeParis}|>]

Out[184]= ClassifierFunction    Method: Markov
                                 Number of classes: 3

In[185]:= a[{Macbeth, AnIdealHusband, TheManWhoLaughs}]

Out[185]= {William Shakespeare, Oscar Wilde, Victor Hugo}
```

As you can see, we have used Markov's method for education, based on which the authors of the works were accurately identified.

Image recognition

With the help of Mathematica's knowledge base on pattern recognition, we can engage in the development of artificial intelligence. Mathematica needs only one photo to report what is depicted in it. This is possible thanks to the ImageIdentify function:

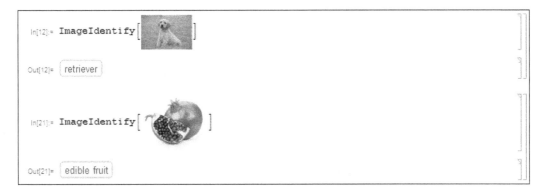

However, this definition is not limited to only one category. Since it is not always possible to identify exactly what is depicted in an image, you should specify additional parameters to be able to select among options. For example, in this case, we are asking for 10 possible options to be shown:

In[23]:= ImageIdentify[🖼 , All, 10]

Out[23]= { edible fruit , pomegranate tree , fruit tree , pomegranate , guava tree , pineapple guava , guava bush , prickly pear , flowering tree , pitahaya }

If we know exactly that there is an edible fruit in the image, then Mathematica will help to classify it. For example, in the next case, we ask for 10 types of edible fruits, which could match this image together with their probabilities:

Note one very useful function—WordCloud. This allows you to build a cloud of words depending on their frequency, which helps to define what is in the image more clearly.

The ImageInstanceQ function is also quite interesting. It allows you to identify whether the object depicted in the image is related to a specific category. For example, we can determine whether there is a tree depicted in the image:

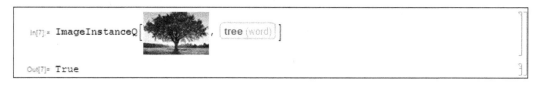

 To enter the name of category—tree—you need to press the *Ctrl* + = key combination and choose the definition you have in mind in the knowledge base.

Recognizing faces

When a photo is added to a social networking website, there is an option to tag friends who are depicted in it. This is done with the help of a face detection function. Using the FindFaces Mathematica function, you can implement a similar feature in your applications. The input parameter of the function is the image and the output parameters are the coordinates of all the rectangles that show the people's faces:

```
In[10]:= pict =            ;

        face = FindFaces[pict]
Out[11]= {{{63.5, 171.5}, {157.5, 265.5}}}

In[13]:= ImageTrim[pict, #] & /@ face

Out[13]= {            }
```

Pay attention to the /@ record, which refers to a short form of the Map[f,expr] function that applies f to each element in the first level of expr. For example, the f /@ {a, b, c, d, e} result is a list of f function applications for each of these parameters: {f[a], f[b], f[c], f[d], f[e]}.

In[57]:= **fam =** **;**

In[58]:= HighlightImage[fam,
    Graphics[{EdgeForm[{Red, Thickness[Large]}], Opacity[0],
      Rectangle @@@ FindFaces[fam]}]]

Out[58]=

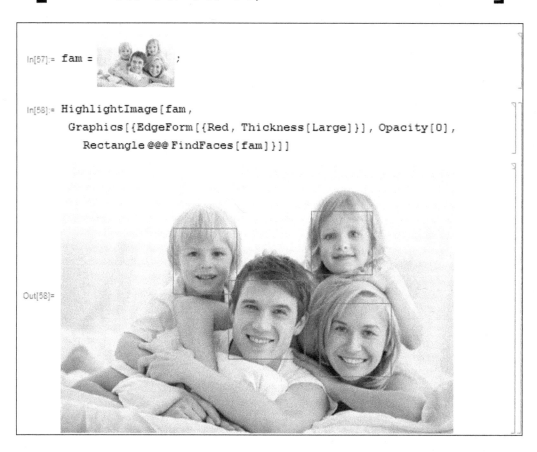

In this example, the FindFaces function has detected multiple faces in a photo and the HighlightImage function made it possible to highlight these faces directly in the photo.

> Take into account the `Rectangle @@@ FindFaces[fam]` record—it is a shortened version of the `Apply[f,expr,{1}]` function that replaces the heads at level 1 of expr by `f`. For example, the result of `f @@@ {{a}, {b}, {c}, {d}}` will be `{f[a], f[b], f[c], f[d]}`.

The `FindFaces` function also has a parameter of the minimum face area size, which helps to prevent incorrect face recognition.

A demonstration of this parameter is shown as follows:

In[61]:= `fam2 =` `;`

In[62]:= `HighlightImage[fam2,`
`    Graphics[{EdgeForm[{Red, Thickness[Large]}], Opacity[0],`
`        Rectangle @@@ FindFaces[fam2]}]]`

Out[62]=

In this picture, we see five areas that were wrongly recognized as faces. We can remove them by limiting the size of the face area from 140 pixels to infinity.

```
In[75]:= HighlightImage[fam2,
    Graphics[{EdgeForm[{Red, Thickness[Large]}], Opacity[0],
        Rectangle @@@ FindFaces[fam2, {140, ∞}]}]]
```

In order to insert the infinity symbol, ∞, press the *Esc* button and type `inf`. Now when you press the *Esc* button again, the infinity symbol will appear.

Recognizing text information

You can solve the problem of textual information recognition using the `TextRecognize` function, which recognizes the text in an image and returns it as a string. This function works with both grey and multichannel images. By default, it recognizes the text written in English; however, using the `Language` parameter, you can specify that the text is written in one of the following languages: French, German, Italian, Portuguese, Russian, and Spanish.

In[103]:= **TextRecognize**[HAVE MORE THAN YOU SHOW, AND SPEAK LESS THAN YOU KNOW.]

Out[103]= HAVE MORE
THAN YOU SHOW,
AND SPEAK LESS
THAN YOU
KNOW.

This function handles different ways to write text, such as text that is written at an angle:

In[104]:= **TextRecognize**[LOST HIGHWAY]

Out[104]= LOST HIGHWAY

In addition to text recognition, you can also speak this text out loud if you choose the Speak representation:

In[105]:= **TextRecognize**[Getting Started with Unity] // Speak

Besides, one of the important functions is determining the language in which the text is written; it helps to perform automatic translation:

In[106]:= **LanguageIdentify**["Слава Україні!"]

Out[106]= Ukrainian

In[108]:= **LanguageIdentify**[**TextRecognize**[Getting Started with Unity]]

Out[108]= English

Recognizing barcodes

The information in the form of a barcode is compact and can be easily recognized by software. In Mathematica, there is a `BarcodeRecognize` function, which can recognize barcodes from an image and provide their meaning. It recognizes 11 one-dimensional barcodes, including the following:

"UPC"	UPC-A	12 numerical digits
"UPCE"	UPC-E	6 numerical digits
"EAN8"	EAN-8	8 numerical digits
"EAN13"	EAN-13	13 numerical digits
"Code39"	Code 39	Up to 43 characters of uppercase letters, numeric digits, and special characters such as -, ., $, /,+, %, and space
"Code93"	Code 93	Uppercase letters, numeric digits, and special characters such as -, ., $, /, +, %, and space
"Code128"	Code 128	Up to 80 ASCII characters
"ITF"	ITF	Up to 80 numerical digits of an even length
"Codabar"	Codabar	Numerical digits and special characters such as :, /, +, .
"GS1"	GS1 DataBar (or RSS)	14 numerical digits
"ExpandedGS1"	GS1 Expanded and Expanded Stacked	74 digits or 41 alphanumeric characters in a single row, or up to 11 stacked rows (GS1 DataBar Expanded Stacked)

The function also recognizes five two-dimensional barcodes, as follows:

{"QR",lev}	QR and error correction level	Variable-length ASCII characters
{"PDF417",lev}	PDF417 and error correction level	Variable-length ASCII characters
"Aztec"	Aztec code	Variable-length ASCII characters
"DataMatrix"	Data Matrix code	Variable-length ASCII characters
"MaxiCode"	MaxiCode	Up to 93 ASCII characters

To get the meaning of the barcode, you need to simply include its image into the `BarcodeRecognize` function parameters:

```
In[110]:= BarcodeRecognize[ ⬛ ]

Out[110]= Analyzing Data with the Help of Mathematica
```

Using additional parameters, you can also find out the type of a barcode:

```
In[111]:= BarcodeRecognize[ ⬛ , {"Data", "Format"}]

Out[111]= {Analyzing Data with the Help of Mathematica, {QR, M}}
```

In the following example, let's write the procedure that decodes the barcode meaning and adds it to the original image:

```
In[120]:= code =         ;

       bc = BarcodeRecognize[code, {"BoundingBox", "Data", "Format"}]

Out[121]= {{{80., 66.}, {1223., 905.}}, 9781782171966, EAN13}

In[123]:= Show[SetAlphaChannel[#[[1]], 0.4],
       Graphics[{EdgeForm[{Thickness[Large], Red}], FaceForm[],
         Rectangle @@ #[[2]],
         Inset[
          Style[#[[3]] <> "\n" <> If[ListQ[#[[4]]], First[#[[4]]],
             #[[4]]], Bold, Black, FontSize → 18],
          RegionCentroid[Rectangle @@ #[[2]]]]}], ImageSize → 200] &[
       Prepend[bc, code]]
```

Let's consider this function. The Show function shows graphics with the specified options added. Using the SetAlphaChannel function, we have set the transparency level of the original image. Then, using the Inset function, we have identified the red-colored area in which the barcode and its text meaning will be highlighted. The boundaries of this area were obtained with the BoundingBox parameter, which we specified when scanning code by the BarcodeRecognize function. Let's recall the #[[1]] record — this means one element of the list, so in our case, #[[1]] is the bc image, #[[2]] indicates the coordinates of the barcode's borders, #[[3]] is the text meaning of the barcode, and #[[4]] is the format of the barcode.

Besides using the BarcodeImage function, you can generate different barcodes by means of software.

Summary

In this chapter, we reviewed the main points of data analysis. We learned Mathematica's functions that will help to perform data classification (as a supervised learning technique) and data clustering (as an unsupervised learning technique). We got to know how to recognize faces, classify objects in an image, and work with textual information by identifying the language of the text and recognizing the text in the image. Apart from this, we analyzed barcodes as a system of information recognition simplification. We learned to read and create barcodes.

In the next chapter, we'll consider how to use Mathematica's tools to analyze different time series consisting of random variables.

5
Discovering the Advanced Capabilities of Time Series

Time series is a very important element of data analysis. When we observe some quantitative phenomenon in time, we witness a time series. This can be the temperature of a place, the population of a country, an exchange rate, the number of students in a classroom, and so on. However, this is not enough just to observe a phenomenon. It is necessary to learn how to forecast and identify trends and the maximum and minimum values within which the time series will exist. Mathematica is of great help in these issues because it contains a lot of functions for time series analysis.

In this chapter, you will learn the following:

- How to set a time series in Mathematica
- How to use the Mathematica time series database in the following fields: weather, astronomy, demography, finance, and so on
- How to work with different models that describe the processes of time series and predict their behavior
- How to conduct tests on autocorrelation, invertibility, and stationarity

Time series in Mathematica

Normally, a time series is set by a number of pairs consisting of the value observation time and the value. For example, in order to monitor the progress in the development of a child, parents measure their height and note the time of the measurement. This data being accumulated can form a time series.

In order to set the time series in Mathematica, you should use the `TimeSeries` function. The plurality of pairs can serve as its parameters:

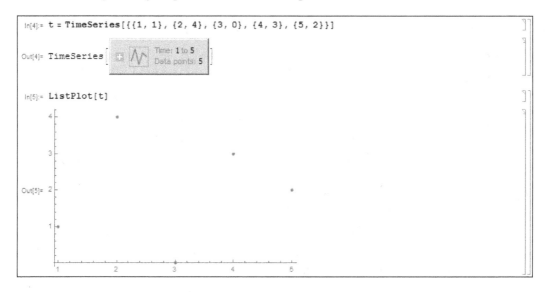

It is also possible to separately specify the set of values and the plurality of time intervals:

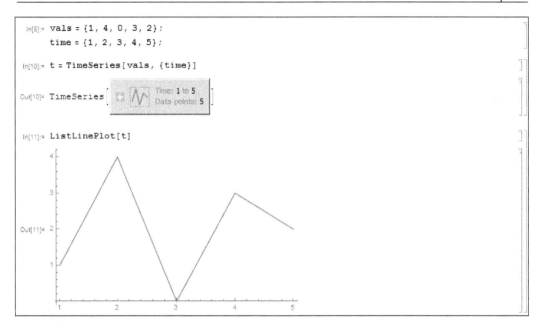

Please note that to build graphs, we've used here different functions: ListPlot for point graphs and ListLinePlot for graphs connecting the points with a line.

If we only have data values that are not tied to a specific interval, then you can use the Automatic parameter that will assign time values:

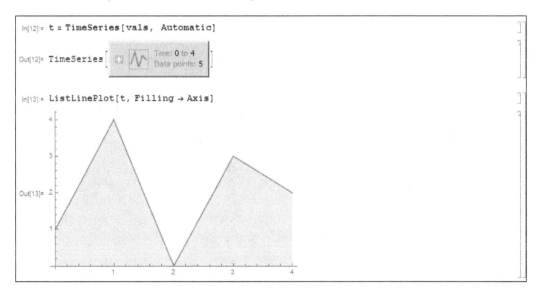

Mathematica can also set a series with only the beginning of a time interval. You can directly indicate that this will be the date:

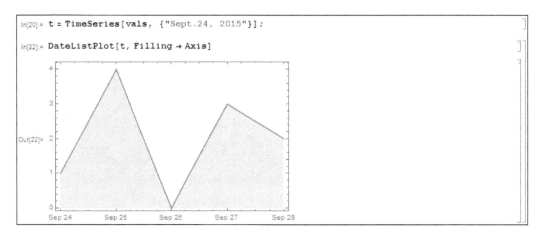

You should pay special attention to the time series in which some data are missing and how they are interpreted. With the `MissingDataMethod` parameter, you can specify that the missing data will be replaced by a constant (the `Constant` value) or will be interpolated (the `Interpolation` value):

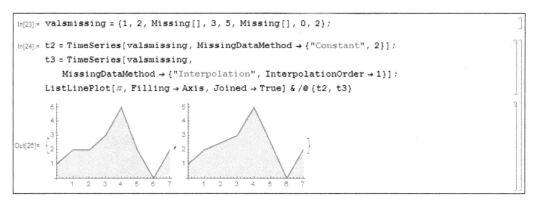

For a time series, you can calculate the average value, the moving average, and the variance (dispersion):

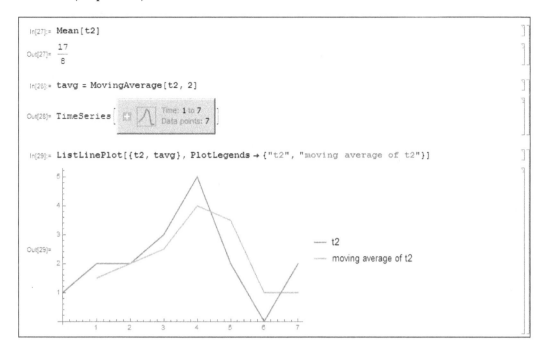

Thus, with the help of Mathematica, you can do the analysis of numerical datasets rather quickly and visually.

Mathematica's information depository

We have considered an elementary example of time series with abstract values; however, in practice, we have to analyze large arrays in order to find patterns and dependencies and to make conclusions from these. In this section, we will review what data has been already collected by Mathematica for our use.

We can take demographic data and use any statistics of the country: GDP, unemployment rate, population, and so on:

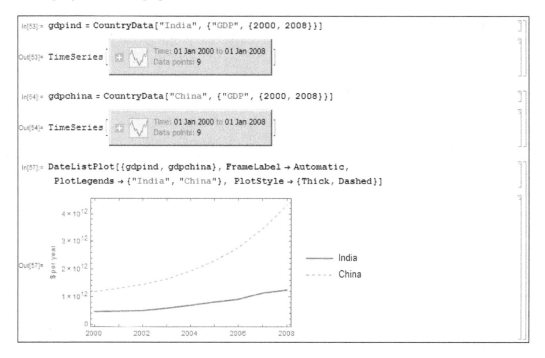

Using the `FinancialData` function, you can access the values of various financial instruments such as share prices and exchange rates.

For example, you can determine the periods of the highest index volatility of Standards & Poor's 500 using the following functions:

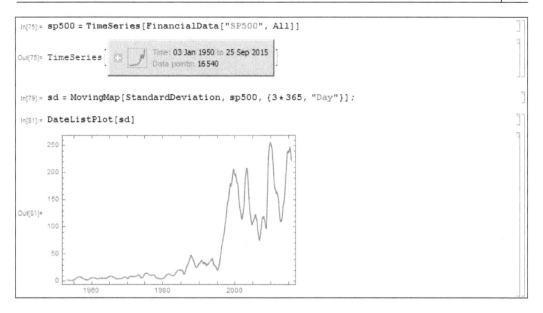

```
In[75]:= sp500 = TimeSeries[FinancialData["SP500", All]]
```

Out[75]= TimeSeries [... Time: 03 Jan 1950 to 25 Sep 2015 Data points: 16540]

```
In[79]:= sd = MovingMap[StandardDeviation, sp500, {3 * 365, "Day"}];
```

```
In[81]:= DateListPlot[sd]
```

In this case, we also got familiar with the MovingMap function that built the time series with the standard deviations of S&P500 index based on the previous 3-year daily data. As you can see, the greatest jump falls during the 2009 crisis.

Using the WeatherData function, we can get the minimum, maximum, and average data on weather conditions for a particular day in a particular region: temperature, pressure, humidity, visibility, wind speed, and so on.

For example, let's see how the maximum and minimum temperatures have varied in London for the past 3 years:

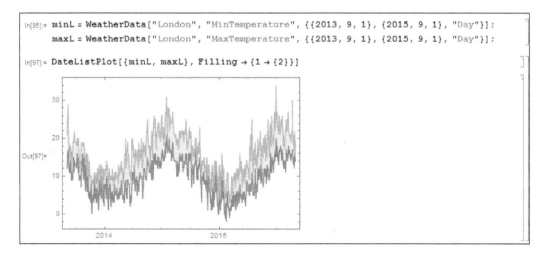

```
In[95]:= minL = WeatherData["London", "MinTemperature", {{2013, 9, 1}, {2015, 9, 1}, "Day"}];
       maxL = WeatherData["London", "MaxTemperature", {{2013, 9, 1}, {2015, 9, 1}, "Day"}];
```

```
In[97]:= DateListPlot[{minL, maxL}, Filling → {1 → {2}}]
```

Here is the list of functions that you can use to obtain additional information:

- `WindSpeedData`: This provides data on wind speed
- `AirPressureData`: This provides pressure data
- `EarthquakeData`: This provides data on earthquakes
- `NuclearReactorData`: This provides data on the properties of various nuclear reactors in the world
- `SunPosition`: This provides data on the position of the sun on a particular day and time
- `MoonPosition`: This provides data on the position of the moon on a given date and time

Also, a whole set of unusual and interesting data can be obtained using the `ExampleData` function. In order to list all the statistics, type `ExampleData["Statistics"]`.

Process models of time series

Since very often we observe random data, in order to predict it, we need to find the most suitable model that would describe the behavior of this data. A time series model that uses random variables is called a process. Thus, if a **time series** is a sequence which is strongly known to us (for example, as a result of observing), then the **time series process** is a random time series, and its values will be different every time depending on the values that the random magnitudes take.

There are several models of time series processes. In order to understand which model is most suitable for sampling data, it is necessary to explore each of them. Next, when we know that the time series refers to a specific type, we can compute the estimates for the model parameters and make a forecast on this basis. For this reason, let's review these models one after another and see how they are implemented in Mathematica.

The moving average model

The moving average model is specified as follows:

$y(t) = c + \left(1 + b_1 \bar{y}(t-1) + \ldots + b_q \bar{y}(t-q)\right) e(t)$, where $y(t)$ is the value of a random process at time, t, c, $b_1, \ldots, b_q$ are constants, $\bar{y}$ is the shift operator, and $e(t)$ is white noise input.

Using the MAProcess function, you can specify such a process. The function parameters are the given constants. For example, let's see how the two implementations of the same process will change in time with these constants: $c=1$, $b_1=2$, and the variation of white noise, 3:

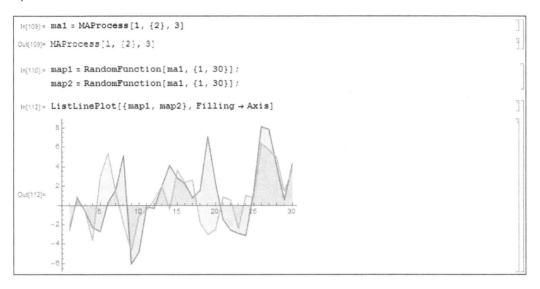

Note that despite the fact that the same RandomFunction[mal, {1, 30}] function has been assigned to map1 and map2, we have got different results. This is due to the fact that the result of the RandomFunction is a random process, as described previously. The other parameter of this function is the time interval from the minimum to maximum value.

The autoregressive process – AR

The autoregressive process is a random process in which the value of a time series at the present moment is linearly dependent on its values in the previous moments. It is described as a solution of the following equation:

$$\left(1 - a_1 E_{t-1} - \ldots - a_p E_{t-p}\right) y(t) = c + e(t)$$, where $c, a_1, \ldots, a_p$ are constants, E is the shift operator, and $e(t)$ is white noise input.

With the `ARProcess` function, you can specify the process. Let's look at its implementation with these constants: $c=0.2$, $a_1=0.5$, $b_2=0.1$ and the variation of white noise, *1*:

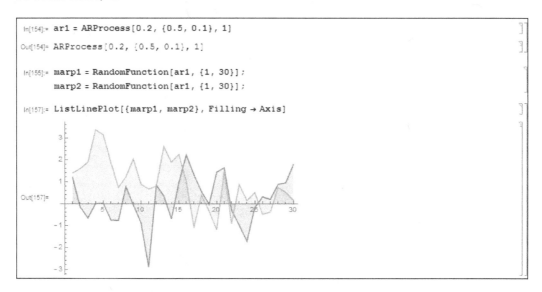

```
In[154]:= ar1 = ARProcess[0.2, {0.5, 0.1}, 1]

Out[154]= ARProcess[0.2, {0.5, 0.1}, 1]

In[155]:= marp1 = RandomFunction[ar1, {1, 30}];
         marp2 = RandomFunction[ar1, {1, 30}];

In[157]:= ListLinePlot[{marp1, marp2}, Filling → Axis]
```

The autoregression model – moving average (ARMA)

The autoregression model combines the previous two models and is the solution of the following equation:

$$\left(1-a_1 E_{t-1}-\ldots-a_p E_{t-p}\right)y(t)=c+\left(1+b_1 E_{t-1}+\ldots+b_q E_{t-q}\right)e(t),$$ where c, $a_1,\ldots,a_p$, $b_1,\ldots,b_q$ are constants, E is the shift operator, and $e(t)$ is white noise.

The first variable of the `ARMAProcess` function is a list of constants for the AR model and the second is a list of constants for the MA model.

Let's consider an example of this function:

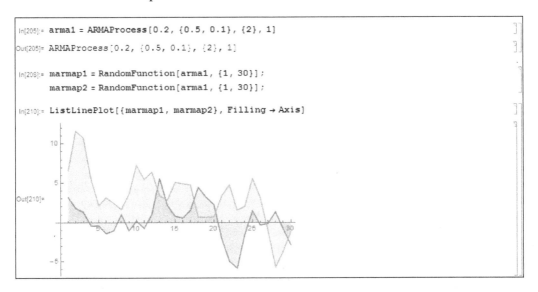

```
In[205]:= arma1 = ARMAProcess[0.2, {0.5, 0.1}, {2}, 1]

Out[205]= ARMAProcess[0.2, {0.5, 0.1}, {2}, 1]

In[206]:= marmap1 = RandomFunction[arma1, {1, 30}];
          marmap2 = RandomFunction[arma1, {1, 30}];

In[210]:= ListLinePlot[{marmap1, marmap2}, Filling → Axis]
```

The seasonal integrated autoregressive moving-average process – SARIMA

If a time series contains a seasonal component, that is, there is a possibility of repeating certain trends at regular intervals, it is convenient to use the SARIMA model. It is described by the following formula:

$$\left(1-a_1 E_{t-1}-\ldots-a_p E_{t-p}\right)\left(1-\alpha_1 E_{t-s}-\ldots-\alpha_m E_{t-ms}\right)\left(1-E_{t-1}\right)^d \left(1-E_{t-s}\right)^\delta y(t) = c + z(t),$$

where $z(t)\left(1+b_1 E_{t-1}+\ldots+b_q E_{t-q}\right)\left(1+\beta_1 E_{t-s}+\ldots+\beta_r E_{t-rs}\right)e(t)$, c, $a_1,\ldots,a_p$, $b_1,\ldots,b_q$, $a_1,\ldots,a_m$, and $\beta_1\ldots\beta_r$ are constants, E is the shift operator, and $e(t)$ is white noise.

In order to set this process, you should call the SARIMA function with the relevant parameters:

SARIMAProcess [{a1, …, ap}, d, {b1, …, bq}, {s, {α1, …, αm}, δ, {β1, …, βr}}, v]

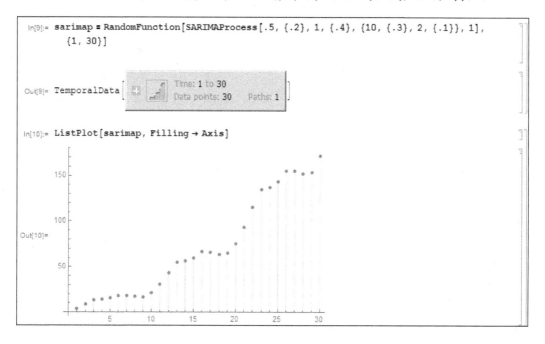

Other time series processes, such as ARCH and GARCH, can also be implemented in Mathematica.

Choosing the best time series process model

As we have become acquainted with the main processes, now we can proceed with their forecasts. To do this, let's get to know the TimeSeriesModelFit function by considering the following example. Suppose we have data on the air temperature in London from January 1, 2012 to September 1, 2015, and we want to get the forecast for the next 6 days. Moreover, we want to compare how the forecast we receive will differ from the actual one. We will also build 95% confidence intervals that should include almost all the possible results of the forecast:

```
In[12]:= weather = TimeSeries[WeatherData["London", "Temperature",
            {{2012, 1, 1}, {2015, 9, 1}, "Day"}, "NonMetricValue"],
           {{2012, 1, 1}, {2015, 9, 1}}];

In[13]:= realweather =
          TimeSeries[WeatherData["London", "Temperature",
             {{2015, 9, 2}, {2015, 9, 7}, "Day"}, "NonMetricValue"],
            {{2015, 9, 2}, {2015, 9, 7}}];

In[14]:= model = TimeSeriesModelFit[weather]
```

Out[14]= TimeSeriesModel[Family: SARIMA   Order: {{6, 0, 0}, {0, 1, 0}₁}]

```
In[17]:= DateListPlot[{TimeSeriesWindow[weather, {{2015, 8, 1}, {2015, 9, 1}}],
            TimeSeriesForecast[model, {6}], realweather},
           PlotLegends → {"weather history", "forecast", "fact"},
           PlotStyle → {Thick, Dashed, Thin}]
```

Out[17]=

Take into account that Mathematica has reported that the SARIMA model is the most suitable one for this time series. This is true, as air temperature is highly dependent on season. Further more, in order to build a graph, we did not take all the data, but just the part from August, by cutting off all other data using the TimeSeriesWindow function. In order to build a forecast, we called the TimeSeriesForecast function, which has generated a sequence of six values based on the model parameters of the SARIMA model. The results' difference with the actual data is just 6° F.

However, be aware that the forecast is just one alternative out of a plurality for the time series development. In order to narrow down the data range, it is necessary to analyze the confidence intervals — that is, such intervals that store 95% of all the forecast data:

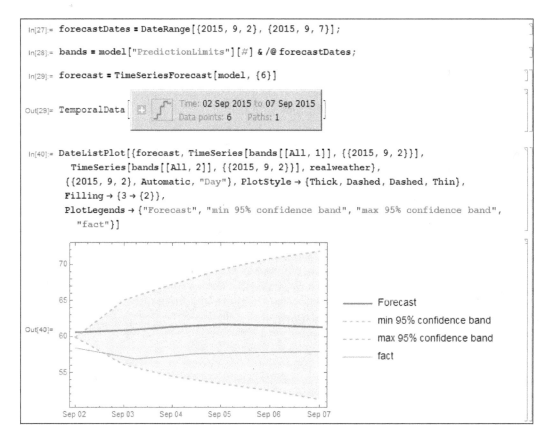

First, we have identified a list of dates for which we want to obtain information about confidence intervals using the DateRange function. Then, we obtained information from our model about the PredictionLimits parameters for this period. To build a graph, we create a time interval with the first and second items of the confidence interval list.

As you can see from the graph, all the data since September 3 (both forecast and actual) fall within our confidence interval.

In order to improve the process model, you can use the AdjustTimeSeriesForecast function that will change the model parameters by considering the new observations:

```
In[9]:=
        proc = model["BestFit"]

Out[9]= SARIMAProcess[0.010019,
        {-0.10824, -0.238406, -0.146326, -0.164096, -0.117813, -0.0452674},
        0, {}, {1, {}, 1, {}}, 5.26892]

In[54]:= sample = RandomFunction[proc, {0, 50}];

In[55]:= forecast = TimeSeriesForecast[proc, sample, {10}];

In[58]:= update = AdjustTimeSeriesForecast[proc, forecast, {15, 16}];

In[60]:= ListLinePlot[{sample, forecast, update},
        PlotLegends → {"data", "forecast", "update"}, PlotStyle → {Thin, Dashed, Thick}]
```

Out[60]=

In this case, we took the model from the previous example and improved it using the conditional data 15 and 16. Note that before the improvement, the trend was down and after the improvement, it went up.

However, the processes are not always suitable for any of the previous models. For example, when we know exactly the basic process model and the only thing that we need is to find out is its parameters. To do this, you can use FindProcessParameters, which allows us to compute the model parameters.

For example, let's generate the Poisson process and check how well this function will compute its parameters:

```
In[42]:= pois = RandomFunction[PoissonProcess[.8], {0, 150}];

In[43]:= param = FindProcessParameters[pois, PoissonProcess[λ]]

Out[43]= {λ → 0.796645}
```

We have generated 150 random values of the time series that satisfy the Poisson law with the $\lambda=0.8$ parameter. After running the `FindProcessParameters` function, whose arguments were our time series and the prospective process, we have obtained an estimate for the λ parameter, which proved to be very close to the initial value.

Tests on stationarity, invertibility, and autocorrelation

When we deal with observed data, we are usually interested in a few things:

- Are we observing a certain constant that simply has some random noisy data? In this case, we check for stationarity (the mean value of the sample does not depend on time).

- Will the process characteristics repeat again after a certain time (invertible processes)?

- Is the observation data dependent on the previous data? Is the seasonality possible (process autocorrelation)?

Let's consider each of these tests in practice.

Checking for stationarity

Here we'll check whether the process is **weakly stationary**. A random process, proc, is weakly stationary if its mean function is independent of time and its covariance function is independent of time translation. This check is done using the `WeakStationarity` function with the random process as its only parameter:

```
In[61]:= WeakStationarity[PoissonProcess[.8]]

Out[61]= False

In[65]:= WeakStationarity[SARIMAProcess[0.01, {-0.1, -0.23}, 0, {0}, {1, {0}, 1, {0}}, 5.26]]

Out[65]= False

In[63]:= WeakStationarity[ARProcess[{a, b, c}, v]]

Out[63]= 1 - c² > 0 && (-b - a c) (b + a c) + (1 - c²)² > 0 &&
          (-(b + a c) (-a - b c) - (-a - b c) (1 - c²)) ((b + a c) (-a - b c) + (-a - b c) (1 - c²)) +
          ((-b - a c) (b + a c) + (1 - c²)²)² > 0
```

Invertibility check

A time series process is **invertible** if it can be written as an autoregressive time series, possibly of infinite order, such that the autoregressive coefficients are absolutely summable. An invertibility check is done using the `TimeSeriesInvertibility` function. This function can be used with time series processes such as MAProcess, ARMAProcess, ARIMAProcess, and FARIMAProcess:

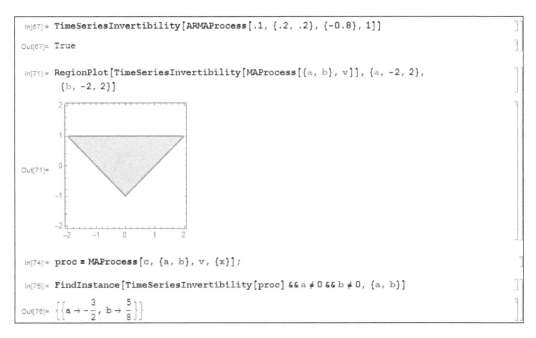

```
In[67]:= TimeSeriesInvertibility[ARMAProcess[.1, {.2, .2}, {-0.8}, 1]]

Out[67]= True

In[71]:= RegionPlot[TimeSeriesInvertibility[MAProcess[{a, b}, v]], {a, -2, 2},
         {b, -2, 2}]
```

```
In[74]:= proc = MAProcess[c, {a, b}, v, {x}];

In[75]:= FindInstance[TimeSeriesInvertibility[proc] && a ≠ 0 && b ≠ 0, {a, b}]

Out[75]= {{a → -3/2, b → 5/8}}
```

In these examples, we have checked the ARMA process for invertibility; using the `RegionPlot` function, we have built the a and b parameters area in which the MA process will be invertible, and we have also found the values of these parameters using the `FindInstance` solution function.

Autocorrelation check

`AutocorrelationTest` performs a hypothesis test for randomness on data with the H_0 null hypothesis that the autocorrelations $\rho_1 = \rho_2 = \ldots = \rho_k = 0$ and alternative H_a that at least one of the $\rho_i \neq 0$. In this way, we are analyzing whether the subsequent data depends on the previous ones.

For simplicity, let's generate a sequence of random variables (some of them are dependent):

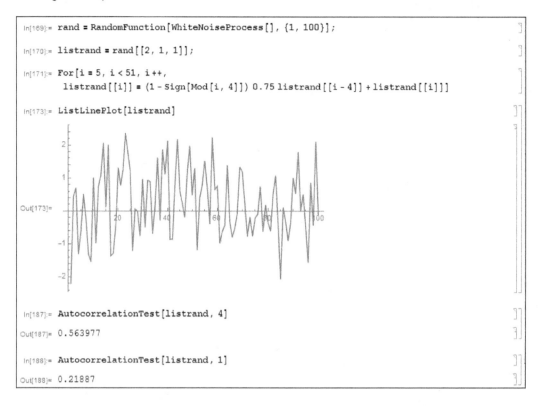

```
In[169]:= rand = RandomFunction[WhiteNoiseProcess[], {1, 100}];

In[170]:= listrand = rand[[2, 1, 1]];

In[171]:= For[i = 5, i < 51, i++,
           listrand[[i]] = (1 - Sign[Mod[i, 4]]) 0.75 listrand[[i - 4]] + listrand[[i]]]

In[173]:= ListLinePlot[listrand]
```

```
In[187]:= AutocorrelationTest[listrand, 4]
Out[187]= 0.563977

In[188]:= AutocorrelationTest[listrand, 1]
Out[188]= 0.21887
```

In this case, the first 100 random white noise values have been generated using the WhiteNoiseProcess function. Then, in the For cycle, we have modified these values so that every fourth value depends on the value that had the 0.75 coefficient 4 moments ago. Thus, we have built a series with an explicit autoregression. After the test, we see that the 4 lag autoregression probability is more than that with 1 lag.

Summary

In this chapter, we became familiar with time series and learned how to process and generate them. We also found out how time series processes are analyzed and what the main model types of these processes are. After knowing the models, we learned how to identify the most appropriate model for a time series, and having determined the model, we could get the forecast data and confidence intervals for future data sets. Using Mathematica functions, we were able to check observation data for stationarity, autocorrelation, and invertibility.

In the next chapter, we will move on to the verification of various statistical hypotheses on the types of sample parameters.

6
Statistical Hypothesis Testing in Two Clicks

When we have data sampling, we are always interested in its characteristics: which law of distribution this data suits the most, what is the mean and variance of the sample, and many other features. However, the verification of these characteristics gives only a probable answer, since we cannot say for sure when we are dealing with random data. There is an entire set of tests that verify this; however, with the help of Mathematica, this check is reduced to two strings of code and you get a quick answer to the question. In this chapter, you will learn the following:

- How to verify that the mean value or the sampling variance is commensurate with a certain quantity
- How to verify that the mean values or the variances of two or more samples are commensurable with each other
- How to test two samples for mutual independence or correlation
- How do I know whether a sample corresponds to a distribution law?

Hypotheses about the mean

In order to check a sample's parameter comparability with certain values, the following hypotheses are suggested: main (zero) hypothesis—when an assigned characteristic is commensurate with this quantity (for example, the mean is 5), and the alternative hypothesis, which differs from the zero hypothesis (for example, the mean is not equal to 5). When we are dealing with a sample of random variable observations, we don't know anything for sure, but we know to a certain degree of probability that this is a degree to which the null hypothesis can be either accepted or rejected. To test a hypothesis, there are functions that calculate the probability of the hypothesis. The more tests there are that give positive results with respect to one hypothesis, the more confidence there is in its truthfulness.

Hypotheses about a mean are checked using the `LocationTest` function. Let's generate a random sample of 1000 values with a normal distribution and test the hypothesis that the sample's mean is 0:

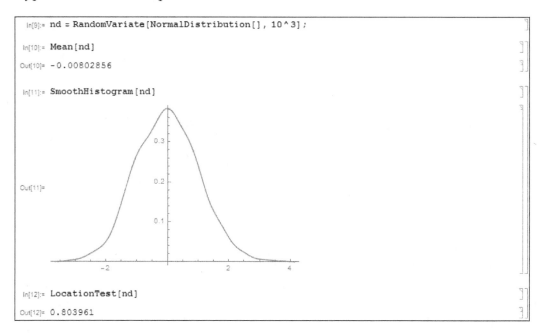

```
In[9]:= nd = RandomVariate[NormalDistribution[], 10^3];
```

```
In[10]:= Mean[nd]
```

```
Out[10]= -0.00802856
```

```
In[11]:= SmoothHistogram[nd]
```

Out[11]=

```
In[12]:= LocationTest[nd]
```

```
Out[12]= 0.803961
```

Note that the random array was generated using `RandomVariate` function. Its first parameter is the type of distribution (in this case, the normal distribution) and the second parameter is the number of values. Here, the mean function has computed the mean of the sample; it turns out to be close to zero. Using the `SmoothHistogram` function, we have plotted a smoothed histogram. As the `LocationTest` function has only one parameter, the sample has issued the probability with 0 as the mean of the sample.

However, in this case, we've got quite a small amount of information. We do not know what tests have been carried out on the sample. It's necessary to specify the name and the results of these tests in scientific works. To list all the tests that have been conducted, you can specify the `TestDataTable` parameter:

```
In[13]:= LocationTest[nd, Automatic, {"TestDataTable", All}]
```

	Statistic	P-Value
Paired T	-0.24829	0.803961
Paired Z	-0.24829	0.80391
Sign	495	0.775964
Signed-Rank	245486.	0.602071
T	-0.24829	0.803961
Z	-0.24829	0.80391

Out[13]=

Let's examine in more detail which tests are conducted with this function:

Test's name	Type of test	Description
"PairedT"	normality	Paired sample test with unknown variance (perform T tests on the paired differences of two datasets)
"PairedZ"	normality	Paired sample test with known variance (perform Z tests on the paired differences of two datasets)
"Sign"	robust	Median test for one sample or matched pairs
"SignedRank"	symmetry	Median test for one sample or matched pairs
"T"	normality	Mean test for one or two samples (performs Student T-test for univariate data and Hotelling's χ^2 test for multivariate data)
"MannWhitney"	symmetry	Median test for two independent samples
"Z"	normality	Mean test with known variance (performs a Z-test assuming the sample variance is the known variance for univariate data, and Hotelling's χ^2 test assuming the sample covariance is the known covariance for multivariate data)

Now, let's test the hypothesis that the sample mean equals 5. To do this, we'll generate two random sequences distributed according to the normal law, but with different characteristics:

```
In[15]:= d1 = RandomVariate[NormalDistribution[5, 2], 1000];
         d2 = RandomVariate[NormalDistribution[1, 1], 1000];

In[17]:= LocationTest[d1, 5]
Out[17]= 0.741254

In[18]:= LocationTest[d2, 5]
Out[18]= 3.33086×10^-165
```

As you can see, the d1 sample mean with the 0.74 probability may be equal to 5. At the same time, the probability that the d2 sample mean is equal to 5 is extremely improbable.

Simultaneously, the null hypothesis can verify that the means of samples d1 and d2 differ by 4:

```
In[27]:= LocationTest[{d1, d2}, 4]

Out[27]= 0.176899

In[28]:= LocationTest[{d1, d2}, 3.9]

Out[28]= 0.973899
```

Note that it turned out that it is most probable that the means differ by 3.9, rather than 4.

If we need to get some properties of the tested hypothesis, it is convenient to do this as follows:

```
In[29]:= d = RandomVariate[NormalDistribution[], {2, 10^3}];

In[30]:= h0 = LocationTest[d, 0, "HypothesisTestData"]

Out[30]= HypothesisTestData        Type: LocationTest
                                    p-Value: 0.93

In[31]:= h0["PValue", "T"]

Out[31]= 0.930127

In[32]:= h0["TestStatistic", "T"]

Out[32]= -0.0876956

In[34]:= h0[{"PValue", "PairedT"}, {"TestStatistic", "PairedT"}]

Out[34]= {0.92916, -0.0889247}
```

In this example, we've generated two arrays of random variables distributed according to standard normal distribution. It was proved that under the T test, the probability that the mean of these samples is 0 is 0.93. According to the `PairedT` test, the probability of a mean of 0 is 0.929 and the test statistic is -0.0889247.

For visual analysis, we can build graphs that would demonstrate the results:

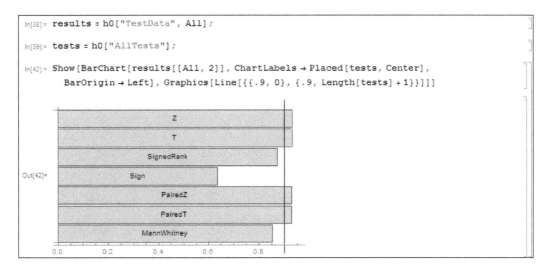

```
In[38]:= results = h0["TestData", All];

In[39]:= tests = h0["AllTests"];

In[42]:= Show[BarChart[results[[All, 2]], ChartLabels → Placed[tests, Center],
          BarOrigin → Left], Graphics[Line[{{.9, 0}, {.9, Length[tests] + 1}}]]]
```

Note that for convenience and with the help of the `Graphics` and `Line` functions, we have drawn the 0.9 level at which the hypothesis is accepted.

Until now, we have tested hypotheses that the mean may be equal to some number; however, you can also test hypotheses that the mean is over/under a certain number. For example, let's test the hypothesis that the mean of our d sample is greater than or equal to -0.2. To do this, let's formulate an alternative hypothesis: the value is strictly less than -0.2:

```
In[47]:= LocationTest[d, -0.2, AlternativeHypothesis → "Less"]

Out[47]= 0.999992
```

Similarly, you can build alternative hypotheses using the values of the `AlternativeHypothesis` parameter: `Unequal`, `Greater`.

It also possible to adjust the significance level of hypotheses and automatically produce a conclusion about the significance of hypotheses:

```
In[48]:= hyp0 = LocationTest[d, .395, "HypothesisTestData", SignificanceLevel → .05];

In[49]:= hyp0["TestConclusion"] // TraditionalForm

Out[49]//TraditionalForm=
       The null hypothesis that the mean difference is 0.395 is rejected at the 5. percent level based on the T test.

In[59]:= hyp1 = LocationTest[d, 0.05, "HypothesisTestData", SignificanceLevel → .05];

In[58]:= hyp1["TestConclusion"] // TraditionalForm

Out[58]//TraditionalForm=
       The null hypothesis that the mean difference is 0.05 is not rejected at the 5. percent level based on the T test.
```

For a statistical data processing example, let's test the hypothesis that male domestic cats have a higher weight than female domestic cats. To do this, let's use the Mathematica database:

```
In[82]:= ExampleData[{"Statistics", "FisherCats"}, "ColumnDescriptions"]

Out[82]= {Sex of the cat "F" or "M", Body weight in kg, Heart weight in g}

In[83]:= {sex, heartW, bodyW} = Transpose@ExampleData[{"Statistics", "FisherCats"}];

In[85]:= LocationTest[{Pick[bodyW, sex, "M"], Pick[bodyW, sex, "F"]}, 0, "TestDataTable",
           AlternativeHypothesis → "Greater"]
```

	Statistic	P-Value
T	6.51786	5.92995×10^{-10}

As you can see, the hypothesis that the weight of a male cat is greater than the weight of a female cat was rejected. Pay attention to the `Pick` function. It allowed us to filter out the input data from a common data set by selecting only the data that belonged to a male or a female.

Let's try to approach the analysis in a different way. Let's take the male cat's and the female cat's heart weight to body weight ratio. We'll test the hypothesis that males and females have the same ratio:

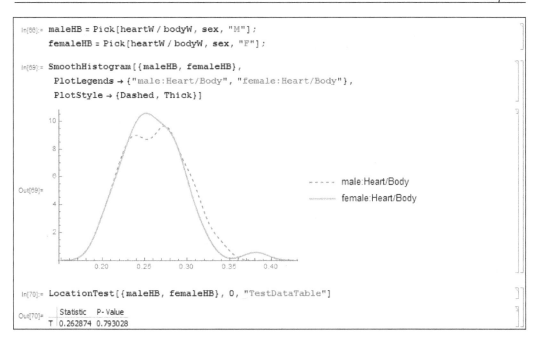

```
In[66]:= maleHB = Pick[heartW / bodyW, sex, "M"];
         femaleHB = Pick[heartW / bodyW, sex, "F"];

In[69]:= SmoothHistogram[{maleHB, femaleHB},
         PlotLegends → {"male:Heart/Body", "female:Heart/Body"},
         PlotStyle → {Dashed, Thick}]
```

```
In[70]:= LocationTest[{maleHB, femaleHB}, 0, "TestDataTable"]
```

	Statistic	P-Value
T	0.262874	0.793028

Out[70]=

Thus our hypothesis can be accepted with the probability of 79.3%.

In order to test the hypothesis on the equality of means of several samples, let's use the `LocationEquivalenceTest` function. It checks the following H_0 hypotheses — the true means of the samples are the same, and H_A — at least one mean differs from the rest. Thus, this function returns the probability that the H_0 hypothesis may be true, that is, the higher the value, the more likely the equality of means will be.

To test these hypotheses, Mathematica proposes conducting the following tests:

Test's name	Type of test	Description
"CompleteBlockF"	Normality, blocked	A mean test for a complete block design (effectively performs a one-way analysis of variance for randomized complete block design)
"FriedmanRank"	Blocked	A median test for a complete block design (ranks observations across rows, and sums the ranks along columns in the data to arrive at the test statistic)

Test's name	Type of test	Description
"KruskalWallis"	Symmetry	A median test for two or more samples (effectively performs a one-way analysis of variance on the ranks of the data)
"KSampleT"	Normality	A mean test for two or more samples (equivalent to a one-way analysis of the variance of the data)

Let's demonstrate an example of how easy it is to test a hypothesis of equality of means:

```
In[73]:= dataStudent = RandomVariate[StudentTDistribution[3], {2, 2000}];

In[75]:= LocationEquivalenceTest[dataStudent, "TestDataTable"]

                        Statistic   P-Value
Out[75]= ─────────────────────────────────────
         Kruskal-Wallis  0.208715   0.647834

In[78]:= LocationEquivalenceTest[dataStudent, "TestConclusion"]

Out[78]= The null hypothesis that the mean difference is 0
         is not rejected at the 5 percent level based on the Kruskal-Wallis test.
```

We have generated two arrays distributed according to a student's distribution, and using the `LocationEquivalenceTest` function, we've tested that the hypothesis on equality of the means doesn't deviate with a 5% level of probability.

Hypotheses about the variance

Suppose we have a sample with σ^2 variance and we assume that the true value of the sample variance is σ_0^2, then we can test the following hypotheses: H_0 - ratio and the alternative hypothesis $H_A : \sigma^2/\sigma_0^2 \neq 1$.

In practice, σ^2 variance is often a measure of risk; therefore, while testing the hypothesis of the equality of variances, we evaluate how accurate is the risk degree is.

The verification of this hypothesis is performed with the help of the `VarianceTest` function. This function can also check hypotheses on the equality of variances in several samples. For example, if one sample's variance is σ_1^2 and the other's - σ_2^2, then the main hypothesis will be an equality check, $\sigma_1^2/\sigma_2^2 = \sigma_0^2$. Mathematica conducts a series of tests to determine the probability of the hypothesis's truth.

Test's name	Type of test	Description
"BrownForsythe"	Robust	A robust Levene test
"Conover"	Symmetry	This is based on squared ranks of the sample data
"FisherRatio"	Normality	This is based on σ_1^2/σ_2^2
"Levene"	robust, symmetry	This compares individual and group variances
"SiegelTukey"	Symmetry	This is based on the ranks of the pooled sample data

Let's generate a random sample and see an example of how this function works:

```
In[2]:= dataVar = RandomVariate[NormalDistribution[], 10^3];

In[6]:= VarianceTest[dataVar]

Out[6]= 0.509212

In[9]:= hyp0 = VarianceTest[dataVar, Automatic, "HypothesisTestData"];

In[11]:= hyp0["TestDataTable"]

Out[11]=        Statistic  P-Value
       Fisher Ratio  1028.12  0.509212
```

Thus, with a probability of 0.509212, the hypothesis on the equality of variance 1 can be accepted. Note the `Automatic` parameter, which selects a value depending on the incoming data array and is equal to 1.

Let's test the hypothesis on the equality of the variances of two samples. In order to do this, let's generate two random arrays distributed according to a student's distribution with a parameter of 5:

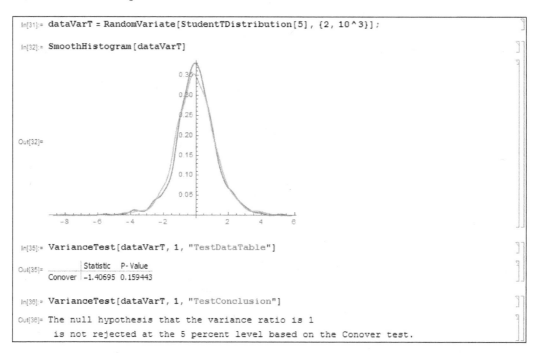

```
In[31]:= dataVarT = RandomVariate[StudentTDistribution[5], {2, 10^3}];
```

```
In[32]:= SmoothHistogram[dataVarT]
```

Out[32]=

```
In[35]:= VarianceTest[dataVarT, 1, "TestDataTable"]
```

Out[35]=

	Statistic	P-Value
Conover	−1.40695	0.159443

```
In[36]:= VarianceTest[dataVarT, 1, "TestConclusion"]
```

Out[36]= The null hypothesis that the variance ratio is 1
is not rejected at the 5 percent level based on the Conover test.

Also, in order to test the hypothesis on equality of variances of two or more samples, we can use the VarianceEquivalenceTest function. Let's generate three sequences of random variables with different means and equal variances:

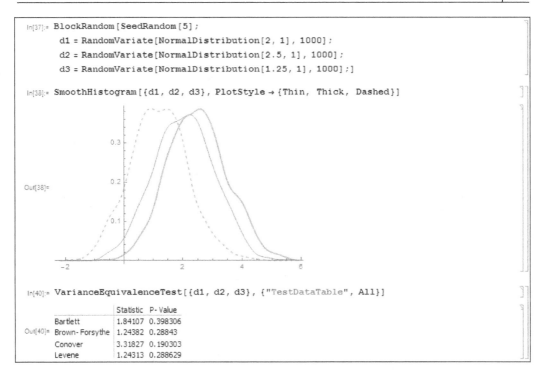

```
In[37]:= BlockRandom[SeedRandom[5];
         d1 = RandomVariate[NormalDistribution[2, 1], 1000];
         d2 = RandomVariate[NormalDistribution[2.5, 1], 1000];
         d3 = RandomVariate[NormalDistribution[1.25, 1], 1000];]

In[38]:= SmoothHistogram[{d1, d2, d3}, PlotStyle → {Thin, Thick, Dashed}]
```

```
In[40]:= VarianceEquivalenceTest[{d1, d2, d3}, {"TestDataTable", All}]
```

	Statistic	P-Value
Bartlett	1.84107	0.398306
Brown-Forsythe	1.24382	0.28843
Conover	3.31827	0.190303
Levene	1.24313	0.288629

Pay attention to the SeedRandom function. It allows you to initiate a sequence of pseudo-random numbers using the initial numbers. This means that you will get exactly the same graphics as a random variables' distribution, despite the fact that the sequence generation took place on independent computers.

Checking the degree of sample dependence

In statistical analysis, it is important to understand whether we are dealing with dependent or independent sets of data. This affects the system model building approach, and thus, the forecast quality.

The independence of the two data samples (represented in a vector or matrix form) is carried out with the help of the IndependenceTest function. This function conducts a series of tests to check the main hypothesis H_0 to see whether the vectors are independent, and to check the alternative hypothesis H_A to see whether the vectors are dependent.

The following tests are conducted:

Test's name	Type of test	Description
"BlomqvistBeta"	Monotonic	This is based on Blomqvist's β
"GoodmanKruskalGamma"	Monotonic, vector	This is based on the γ-coefficient
"HoeffdingD"	Vector	This is based on Hoeffding's D
"KendallTau"	Monotonic	This is based on Kendall's τ-b
"PearsonCorrelation"	Linear, normality, vector	This is based on Pearson's product-moment r
"PillaiTrace"	Normality, linear	This is based on Pillai's trace
"SpearmanRank"	Monotonic	This is based on Spearman's ρ
"WilksW"	Normality, linear	This is based on Wilks' W

This function returns the probability with which the hypothesis H_0 can be accepted. Thus, the smaller this value is, the more likely it will be that the hypothesis is false.

Let's consider a function example. To do this, we'll generate two arrays of random variables distributed according to Binormal Distribution:

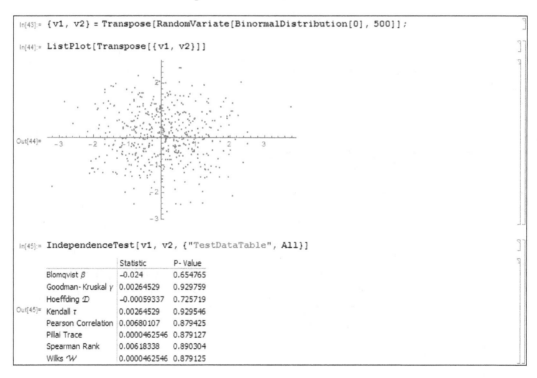

In[43]:= `{v1, v2} = Transpose[RandomVariate[BinormalDistribution[0], 500]];`

In[44]:= `ListPlot[Transpose[{v1, v2}]]`

In[45]:= `IndependenceTest[v1, v2, {"TestDataTable", All}]`

	Statistic	P-Value
Blomqvist β	-0.024	0.654765
Goodman-Kruskal γ	0.00264529	0.929759
Hoeffding D	-0.00059337	0.725719
Kendall τ	0.00264529	0.929546
Pearson Correlation	0.00680107	0.879425
Pillai Trace	0.0000462546	0.879127
Spearman Rank	0.00618338	0.890304
Wilks W	0.0000462546	0.879125

As you can see, most of the tests have accepted the main hypothesis on the independence of the data samples with a sufficiently high degree of probability.

Let's get acquainted with the additional parameters of the test. To do this, we need to get access to the `HypothesisTestData` object:

```
In[46]:= h0 = IndependenceTest[v1, v2, "HypothesisTestData"]

Out[46]= HypothesisTestData     Type: IndependenceTest
                                 p-Value: 0.879

In[47]:= h0["DegreesOfFreedom"]
Out[47]= 498

In[48]:= h0["ShortTestConclusion"]
Out[48]= Do not reject

In[49]:= h0["TestConclusion"]
Out[49]= The null hypothesis that the populations are independent
         is not rejected at the 5 percent level based on the Pearson Correlation test.
```

In order to adjust the hypothesis' deviation, you should use the `SignificanceLevel` parameter. For example, in our case, we'll set this level higher:

```
In[62]:= IndependenceTest[v1, v2, "TestConclusion", SignificanceLevel → .25] //
           TraditionalForm

Out[62]//TraditionalForm=
         The null hypothesis that the populations are independent
         is not rejected at the 25. percent level based on the Pearson Correlation test.
```

But what if the two samples prove to be dependent? In this case, you need to determine how dependent on each other they are, or in other words, how correlated they are. If the correlation coefficient of the two samples is equal to zero, this indicates their potential independence from each other. And if it's 1 or -1, then that's a potential linear dependence.

Checking the main hypothesis (that the two-dimensional sample has a correlation coefficient of ρ_0) is done by the `CorrelationTest` function.

The following tests are used to check this hypothesis:

Test's name	Description
"PearsonCorrelation"	This is based on the Pearson product-moment r (this assumes that the data was drawn from a normal distribution)
"SpearmanRank"	This is based on Spearman's ρ

Here is a simple example of when there is independent data:

```
In[64]:= m = RandomVariate[BinormalDistribution[0], 1000];

In[65]:= CorrelationTest[m, 0, {"TestDataTable", All}]
```

	Statistic	P-Value
Out[65]= Pearson Correlation	−0.0204636	0.518127
Spearman Rank	−0.0267615	0.397997

```
In[66]:= Correlation[m][[1, 2]]

Out[66]= -0.0204636
```

Note that the correlation coefficient between the elements of the sample is -0.0204636, and the hypothesis that the correlation coefficient is zero can be accepted with a probability of 0.518127 according to Pearson's test and 0.397997 according to Spearman's rank.

Let's check the dependence between body weight and heart weight of domestic cats:

```
In[89]:= {sex, heartW, bodyW} = Transpose@ExampleData[{"Statistics", "FisherCats"}];

In[105]:= Correlation[Transpose[{bodyW, heartW}]][[1, 2]]

Out[105]= 0.804127

In[106]:= CorrelationTest[Transpose[{bodyW, heartW}], 0.8, {"TestDataTable", All}]
```

	Statistic	P-Value
Out[106]= Spearman Rank	0.790843	0.741931

```
In[109]:= CorrelationTest[Transpose[{bodyW, heartW}], 0.99, {"TestDataTable", All}]
```

	Statistic	P-Value
Out[109]= Spearman Rank	0.790843	3.45658×10^{-78}

As you can see, the dependence is very significant — the correlation coefficient with a high probability is 0.790843, and at the same time, the hypothesis on a linear dependence was rejected. Pay attention to the `Transpose` function. With its help, we have created a two-dimensional array out of two data vectors.

Hypotheses on true sample distribution

In order to forecast the behavior of the data sample, you need to know not only its parameters, such as mean and variance, but also the distribution law, which controls the data. There are many distribution laws, and to suggest a hypothesis on similarity, you need to know the unique characteristics of each distribution. It is often sufficient to study a sample histogram to make a choice.

Using the `DistributionFitTest` function, you can test the hypothesis that the dataset was drawn from a population with a distribution, and the alternative hypothesis H_A that it was not.

In order to check the main hypothesis, the data sample is tested. This tests the mean assessment of the difference $d(x)$ of the empirical value of the distribution function $\hat{F}(x)$ and its predicted value, $F(x)$. The following tests are conducted for univariate or multivariate distributions:

Test's name	Type of test	Description
"AndersonDarling"	Distribution, data	This is based on Expectation$[\dfrac{d(x)^2}{F(x)(1-F(x))}]$
"CramerVonMises"	Distribution, data	This is based on Expectation[d(x)2]
"JarqueBeraALM"	Normality	This is based on skewness and kurtosis
"KolmogorovSmirnov"	Distribution, data	This is based on $\sup_x |d(x)|$
"Kuiper"	Distribution, data	This is based on $\sup_{d(x)>0} d(x) - \inf_{d(x)<0} d(x)$
"PearsonChiSquare"	Distribution, data	This is based on the expected and observed histogram
"ShapiroWilk"	Normality	This is based on quantiles
"WatsonUSquare"	Distribution, data	This is based on Expectation$[(d(x)-\bar{d}(x))^2]$

The following tests are for for multivariate distributions:

Test's name	Type of test	Description
"BaringhausHenze"	Normality	This is based on the empirical characteristic function
"DistanceToBoundary"	Uniformity	This is based on the distance to uniform boundaries
"MardiaCombined"	Normality	This combines Mardia skewness and kurtosis
"MardiaKurtosis"	Normality	This is based on multivariate kurtosis
"MardiaSkewness"	Normality	This is based on multivariate skewness
"SzekelyEnergy"	Data	This is based on Newton's potential energy

To see an example of this function's use, let's determine which distribution is best suited for day-to-day point changes in the S&P 500 index.

Firstly, using the Last function, let's select all the values of the S&P 500 and determine the change degree at the logarithmic scale:

```
In[118]:= sp500 = Last /@ FinancialData["SP500", All];

In[119]:= d = Log[Most[sp500] / Rest[sp500]]

         {-0.01134, -0.00473654, -0.00294898, -0.00587201,
          0.00293163, -0.00351694,            , 0.00359469, -0.00800352,
Out[119]=  -0.00877978, -0.000724849, -0.00127466, 0.00684882}
```

For more clarity, we'll represent the data by frequency in the form of a histogram by limiting its range of values from -0.05 to 0.05:

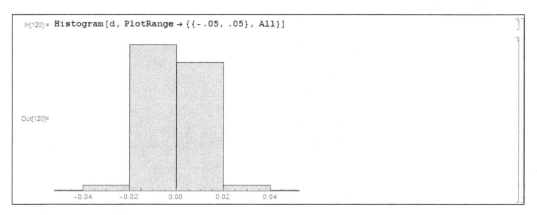

```
In[120]:= Histogram[d, PlotRange → {{-.05, .05}, All}]



Out[120]=
```

As you can see, the histogram has a symmetrical shape, so you should search among distributions with a symmetrical density. A similar distribution to this might be the Laplace distribution. Let's look at its distribution density function:

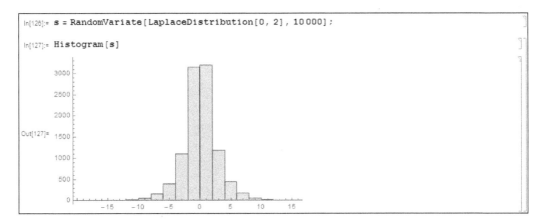

Let's test the hypothesis that the index changes logarithm may be distributed under this law:

```
In[135]:= h0 = DistributionFitTest[d, LaplaceDistribution[a, b], "HypothesisTestData"];
```

```
In[136]:= h0["TestDataTable", All]
```

	Statistic	P-Value
Anderson-Darling	5.68806	1.08522×10^{-6}
Cramér-von Mises	0.650543	0.
Pearson χ^2	212.507	5.60913×10^{-11}

```
In[137]:= h0["FittedDistribution"]
```

```
Out[137]= LaplaceDistribution[-0.000468177, 0.00654674]
```

```
In[138]:= h0["TestConclusion"] // TraditionalForm
```

Out[138]//TraditionalForm=

The null hypothesis that the data is distributed according to the LaplaceDistribution[a, b] is rejected at the 5 percent level based on the Cramér-von Mises test.

As you can see, the hypothesis was rejected because of a too high deviation statistics. However, the most eligible Laplace distribution parameters were obtained for a given sample. Let's see how visually different the forecast data is compared to the real data within the interval from -0.1 to 0.1:

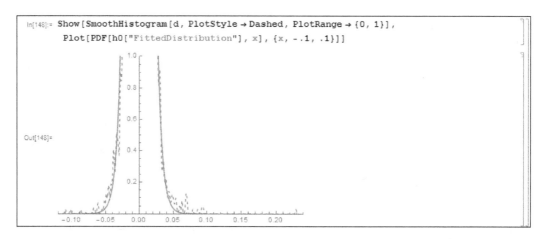

```
In[148]:= Show[SmoothHistogram[d, PlotStyle → Dashed, PlotRange → {0, 1}],
          Plot[PDF[h0["FittedDistribution"], x], {x, -.1, .1}]]
```

As you can see, small data deviations fit the most into this model. It should also be noted that we took the raw data "as is", without prefiltration of large bursts, which significantly affected the test results.

Pay attention to the PDF function that allows you to get a probability density function.

Summary

In this chapter, we learned how to test hypotheses on possible parameters of a sample and also received evidence that this task does not take much time. We learned that in order to estimate a sample mean, we can use the LocationTest function, and with its help, we can test hypotheses on the equality of the means of two samples. We also acquired a skill to test hypotheses on the true value of a variance with the help of the VarianceTest function. We found out that we can check the degree of dependence of data samples using the CorrelationTest function. In the end, we learned how to test hypotheses on the true distribution of a sample with the help of the DistributionFitTest function. In the next chapter, we will learn how to predict data using all the knowledge we've gained so far.

7
Predicting the Dataset Behavior

Analyzing data is important to identify and understand inner dependencies. By examining the exchange rates statistics, we can trace how they grow or fall depending on the political, economic, psychological, and even informational events. These dependencies will subsequently help to determine the future rate value in case of repeated factors influence. So, data analysis often comes down to learning how to make predictions. However, you can predict not just numerical data. Mathematica has gone beyond the number fields long ago and handles audio, video, and text information as easily as numbers.

From this chapter, you will obtain the necessary knowledge to learn how to predict with Mathematica, such as the following:

- Predicting the future values of sampling data
- Intelligent picture processing and replicating style patterns
- Predicting with the help of probability automaton model

Classical predicting

In the previous chapters, we became familiar with data samples and time series and got to know how to define their parameters, which means we were able to predict future values. As a matter of fact, this is classical prediction. However, if the statistical tools seems a bit difficult for you and there is no time to gain an understanding, you can use a quicker solution—the Predict function. After receiving an input data array, it immediately issues a predicted value by keeping all the calculations behind the scenes:

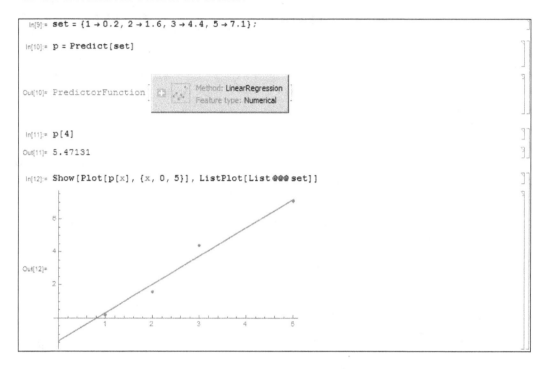

In this case, we took a preliminary dataset—note the list entry in the format: input data -> value. Then, using the Predict function, we obtained PredictorFunction that can output any prediction value depending on the input data. For example, if the input value is equal to 4, the output will be 5.47. After reviewing our data, Mathematica came to the conclusion that the best model for prediction is linear regression. With the graph that we have built by successively substituting values from 0 to 5, you can see how little the difference is between our actual values and those predicted. For more information about the model parameters, use the PredictorInformation function:

```
In[13]:= PredictorInformation[p]
```

Predictor information

Method	Linear regression
Number of features	1
Number of training examples	4
L1 regularization coefficient	0
L2 regularization coefficient	0.1

Out[13]=

However, you can directly specify the method of prediction. There are four methods:

LinearRegression	This predicts from the linear combinations of features
NearestNeighbors	This predicts from the nearest neighboring examples
NeuralNetwork	This predicts using an artificial neural network
RandomForest	This predicts from the Breiman–Cutler ensembles of decision trees

For example, let's take a sample where the outcome depends nonlinearly on the input data and apply two different methods:

```
In[14]:= d = {1, 2, 3, 4, 5, 6, 7, 8, 9} → {1, 2, 3, 4, 5, 6, 7, 8, 9}^3;
```

```
In[15]:= {neur, line} = Predict[d, Method → #] & /@ {"NeuralNetwork", "LinearRegression"}
```

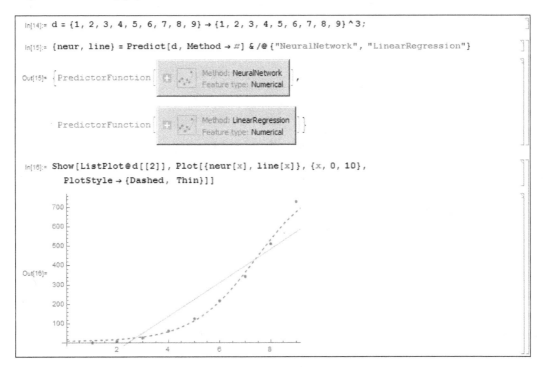

```
In[16]:= Show[ListPlot@d[[2]], Plot[{neur[x], line[x]}, {x, 0, 10},
          PlotStyle → {Dashed, Thin}]]
```

As you can see, the prediction built with the help of a neural system gave much better results than linear regression.

Let's consider a case where it is necessary to predict the initial value based on multiple input values. Since the wine quality evaluation depends on many characteristics such as acidity, density, and the level of alcohol, we can determine the level of appraisal for a certain wine with excellent characteristics by knowing the estimates of the wine from experts. To do this, let's use the `WineQuality` data from the `MachineLearning` library:

```
In[37]:= ExampleData[{"MachineLearning", "WineQuality"}, "LongDescription"]

Out[37]=      Predict the subjectively reported quality of a white
         wine (on a scale of 1-10), given 11 physical features of the wine.
            These features include properties like the pH of the
         wine and its alcohol content.There are 4898 examples.

In[38]:= ExampleData[{"MachineLearning", "WineQuality"}, "VariableDescriptions"]

Out[38]= {fixed acidity, volatile acidity, citric acid, residual sugar,
         chlorides, free sulfur dioxide, total sulfur dioxide, density,
         pH, sulphates, alcohol} → wine quality (score between 1-10)
```

Let's create data for training and see the boundaries of wine characteristics such as the residual sugar and alcohol:

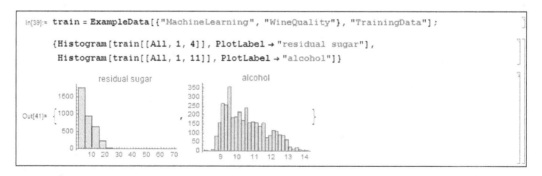

```
In[39]:= train = ExampleData[{"MachineLearning", "WineQuality"}, "TrainingData"];

      {Histogram[train[[All, 1, 4]], PlotLabel → "residual sugar"],
       Histogram[train[[All, 1, 11]], PlotLabel → "alcohol"]}
```

We'll provide the data for training to the input of the `Predict` function and see how it will rate any random wine:

```
In[42]:= p = Predict[train]

Out[42]= PredictorFunction [ ⊕ ⌁⌁ ]  Method: RandomForest
                                      Number of features: 11

In[43]:= uwine = {5.4`, 0.34`, 0.36`, 8.9`, 0.041`, 5.8`, 181.`, 0.99512`, 2.02`, 0.65`, 9.1`};

In[44]:= p[uwine]

Out[44]= 5.03925
```

Now we can select the optimal values for the residual sugar and alcohol characteristics that would guarantee a better appraisal for this wine. To do this, let's just build a three-dimensional graph of a prediction function based on these parameters:

```
In[45]:= quality[sugar_, alcohol_] :=
           p[{5.4`, 0.34`, 0.36`, sugar, 0.041`, 5.8`, 181.`, 0.99512`, 2.02`, 0.65`, alcohol}];

In[46]:= Show[Plot3D[quality[sugar, alcohol], {sugar, 0, 20}, {alcohol, 8, 14},
           AxesLabel → Automatic], ListPointPlot3D[{{8.9`, 9.1`, p[uwine]}},
           PlotStyle → {Black, PointSize[.05]}]]
```

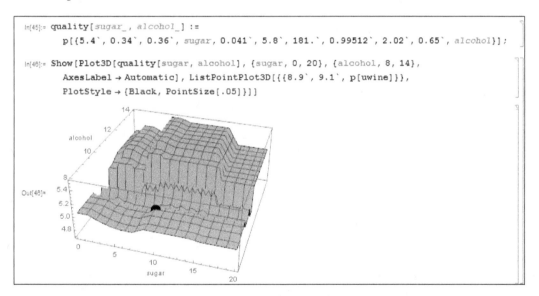

The initial state of the residual sugar and alcohol characteristics on the graph is indicated in black. Having all other parameters constant, we can see that the wine appraisal level can be improved to 5.6 by slightly reducing the sugar characteristic and increasing the alcohol characteristic.

If some data is missing, the prediction can still be performed by substituting the data with the `Missing` function:

```
In[47]:= p = Predict[{{1.8, "male"} → 1.2, {3.1, Missing[]} → 1.5, {Missing[], "female"} → 6.1,
         {3.2, "female"} → 9, {Missing[], "male"} → 0.8, {1.9, "male"} → 4},
         Method → "NearestNeighbors"]

Out[47]= PredictorFunction    Method: NearestNeighbors
                              Number of features: 2

In[49]:= p[{{4, "male"}, {Missing[], "female"}, {9.5, Missing[]}, {Missing[], Missing[]}}]

Out[49]= {1.5, 6.1, 9., 0.8}
```

As you can see, despite the complete absence of input data, the prediction function has still given an outstanding result.

The `Predict` function also works with graphics that greatly facilitates data processing. For example, let's choose the graphical information as an input array, and the result will be the shaded data area:

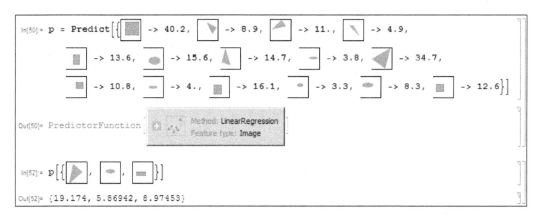

As you can see, the `Predict` function handled the task pretty well—the initial values as well as the areas correlate with each other.

Image processing

One of the unique features of predicting is the obtaining of similar data. For example, a comic impersonator after having learned the peculiarities of a famous person's voice begins to speak with their intonation. In this section, we will see how Mathematica, after having learned the features of an artist's style, can continue his painting. This opens up new possibilities in the field of data restoration.

For example, we'll take Claude Monet's painting *Water Lilies*:

```
In[1]:= monet =
    Import[
      "https://upload.wikimedia.org/wikipedia/commons/3/35/Monet_-_Seerosen_1906.jpg"],
    monet = ImagePad[monet, -50];
```

Using the `ImagePad` function, we cut it off on all sides by 50 pixels. Then, we process the image to enable Mathematica to continue this:

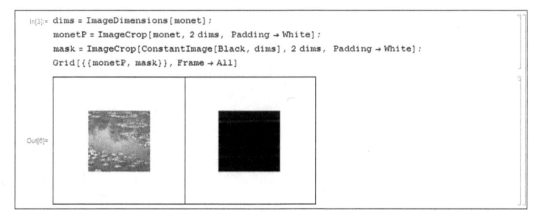

```
In[3]:= dims = ImageDimensions[monet];
    monetP = ImageCrop[monet, 2 dims, Padding → White];
    mask = ImageCrop[ConstantImage[Black, dims], 2 dims, Padding → White];
    Grid[{{monetP, mask}}, Frame → All]
```

With the help of the `ImageDimensions` function, we have got an array consisting of the length and width values of the image. Then, using the `ImageCrop` function, we have created a new image, which is two times bigger than the previous one. At the same time, we have created an image of exactly the same size that will be used as a mask for further calculations. Note that the black rectangle was created using the `ConstantImage` function.

In the next stage, we'll directly deal with the completion of the image:

```
In[7]:= HighlightImage[Inpaint[monetP, mask, Method → "TextureSynthesis"],
    Dilation[MorphologicalPerimeter[mask], 5], "HighlightColor" → White,
    Method → "Solid"]
```

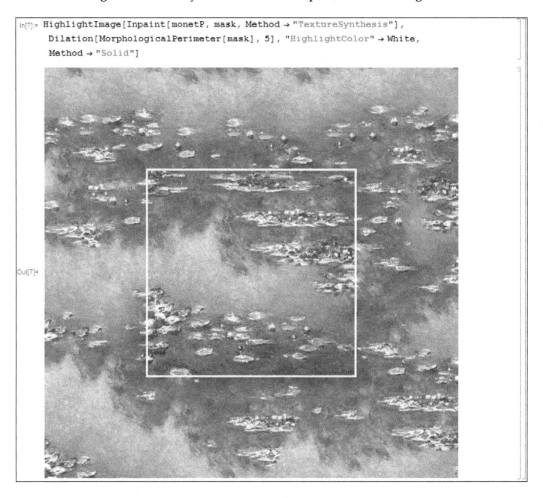

Out[7]=

The central magic function that has performed the image completion is `Inpaint`. Let's examine it in detail. The first parameter of the function is all future image area and the second is the masked area, where the black color indicates what remains intact and white indicates what should be restored. The third selected parameter is the `TextureSynthesis` method. This method ensures the search for the most suitable image texture. The white frame at the center of the image shows that the original was obtained with the help of `Dilation[MorphologicalPerimeter[mask], 5]`.

To get more realistic results, you need to select other parameter values of the Inpaint function: NeighborCount is the number of nearby pixels used for texture comparison and MaxSamples is the maximum number of samples used to find the best-fit texture.

Look and compare the result during the selection of the NeighborCount =1000 and MaxSamples=40 parameters.

```
In[53]:= HighlightImage[
    Inpaint[monetP, mask,
      Method → {"TextureSynthesis", "NeighborCount" → 1000, "MaxSamples" → 40}],
    Dilation[MorphologicalPerimeter[mask], 5], "HighlightColor" → Brown,
    Method → "Solid"]
```

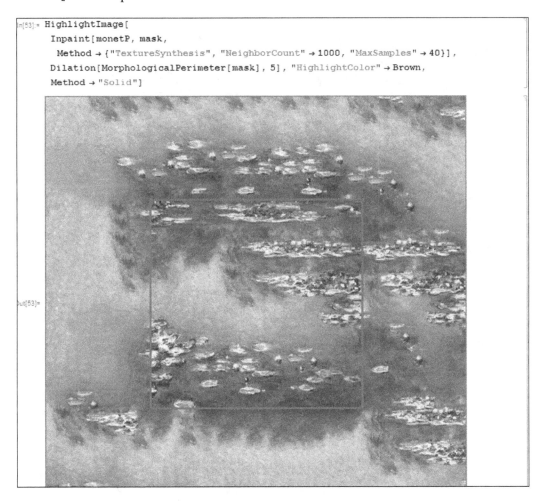

Out[53]=

In this case, the number of water lilies have been reduced and they have become more homotypic than in the previous example.

The main intelligent components of the `Inpaint` function is assistance in photo retouching and restoration of the missing parts. For example, let's take a picture with one end cut off:

Let's perform an intelligent restoration of this picture. We will remove all the cars and the black stripe on the right side while maintaining the general composition. For this purpose, we'll create a filter — a black image with white-colored areas that indicate the parts needed to be removed from the picture.

Then, let's call the Inpaint function indicating the original image as the first parameter and the mask as the second parameter:

Having compared the two images, you can get evidence that the unnecessary objects were removed without any loss of quality.

Probability automaton modelling

In order to predict the behavior of more complex systems that are affected by random factors, you can use modeling. For example, the **probability automaton modeling method** allows us to make a model in the form of interconnected automatons whose states are changing in time simultaneously and discretely. The automaton receives some input signal, has an internal state that includes the state generated by a random value, and is capable of producing an output signal.

Let's consider a simple model of an ATM and its implementation in Mathematica. Let's assume that there is an ATM, which is replenished every t interval by a constant r value. The ATM is approached at random intervals by customers, who withdraw a random amount of money. If there is no money, the ATM receives a negative feedback with a value, v, and to deliver the money, you need to spend certain variables. At the same time, the bank receives income for every unit of time that amounts to an α percent of the investment of the clients' temporarily free funds being stored in the ATM. The task is to predict the profit from the ATM in N timeslots.

In order to assign the probability automaton model, it is necessary to determine its components: the rules of the automatons' internal state changes, the automatons' outputs, and the distribution laws of random variables.

 More details on this method can be found in the monograph *Algorithmische Modellierung Ökonomischer Probleme*, Bakaev, A., N.I. Kostina und N.W. Jarowizki, Editorial: Berlin, Akademie Verlag, 1974

Let's denote the internal states of the automatons as follows:

- A1: The time before the arrival of the customer to withdraw money
- A2: The time before the arrival of the collector car
- A3: The amount withdrawn by the client
- A4: The money in the ATM
- A5: The profit of the ATM

The following are the rules of the internal state changes:

- $a1(t+1) = \text{if } a1(t)>0 \text{ then } a1(t)-1 \text{ else } \xi$
- $a2(t+1) = \text{if } a2(t)>0 \text{ then } a2(t)-1 \text{ else } t$
- $a3(t+1) = \eta$
- $a4(t+1) = \max\{0, \ a4(t)-x1(t)a3(t)+x2(t)r\}$
- $a5(t+1) = \max\{0, \ a5(t)-x1(t)(1-x4(t))v-x2(t)s+\alpha a4(t)\}$

Along with the conditions of transition from one internal state to another, let's assign how the output signals of the five automatons will look:

- `x1(t)=if a1(t)>0` then 0 else 1

- `x2(t)=if a2(t)>0` then 0 else 1

- `x3(t)=a3(t)`

- `x4(t)=if a4(t)>0` then 1 else 0

- `x5(t)=a5(t)`

Let's explain some of these, as follows:

The output signal of the `A1` automaton is equal to 1 at the time when a customer approaches the ATM to withdraw cash and 0 in other cases.

The output signal of the `A2` automaton is equal to 1 at the time when the ATM is loaded with cash.

The output signal of the `A4` automaton is equal to 1 when there is cash in the ATM and 0 otherwise.

To specify this model in Mathematica also, let's assume that the ξ random variable is distributed according to the Poisson law with parameter 5, and the η random variable is distributed according to the normal law with the parameters 50,000 and 20,000. Let's create the `atm` function, which will perform model computing:

```
In[102]:=
    atm[t_, s_, r_, a_, n_, v_] :=
    Block[{i = 1, ξ = RandomVariate[PoissonDistribution[5], n],
        η = RandomVariate[NormalDistribution[50000, 2000], n], a = Array[0 &, {5, n}],
        x = Array[0 &, {5, n}]}, a[[1, 1]] = 5;
    a[[2, 1]] = t;
    a[[3, 1]] = 1000;
    a[[4, 1]] = 100000;
    a[[5, 1]] = 0;
    For[i = 1, i < n, i++, x[[1, i]] = If[a[[1, i]] > 0, 0, 1];
        x[[2, i]] = If[a[[2, i]] > 0, 0, 1];
        x[[3, i]] = a[[3, i]];
        x[[4, i]] = If[a[[4, i]] > 0, 1, 0];
        x[[5, i]] = a[[5, i]];
        a[[1, i + 1]] = If[a[[1, i]] > 0, a[[1, i]] - 1, ξ[[i + 1]]];
        a[[2, i + 1]] = If[a[[2, i]] > 0, a[[2, i]] - 1, t];
        a[[3, i + 1]] = η[[i + 1]];
        a[[4, i + 1]] = Max[0, a[[4, i]] - x[[1, i]] a[[3, i]] + x[[2, i]] r];
        a[[5, i + 1]] = a[[5, i]] - x[[1, i]] (1 - x[[4, i]]) v - x[[2, i]] s + a a[[4, i]];];
    List[a, x]]
```

As the function's parameters, we have chosen all the main constants of the model. Then, we have identified the main variables and generated the required number of random variables. We have also specified the initial states of the automatons with which the computing will be continued. In the cycle, we have successively computed the output signals and the internal states of each automaton.

Let's consider the model's specific input data and look at its behavior:

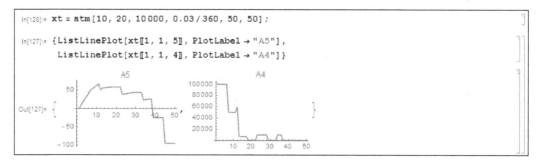

As you can see, after the 20th point in time, the ATM contained quite a small amount of cash and the collectors didn't manage to replenish it in full. As a result, the ATM started being unprofitable, and after the 40th point in time, its replenishment became simply unremunerative.

Similarly, you can describe and formalize the complex system by studying the internal connections and random variables. Using simulation, this approach provides a prediction that is impossible to obtain with ordinary predictions.

Summary

In this chapter, we considered how to find regularities and predict the behavior of numeric data with the help of Mathematica. We got to know which parameters of the `Predict` function can improve the quality of a prediction. We also became familiar with the possibilities of intelligent processing of graphical information using the `Inpaint` function and learned to imitate an author's style by expanding their work or restoring it. Using the methodology of probability automaton modeling, we were able to build a model of a complex system to build a prediction with the parameters of a system.

In the next chapter, we will apply most of our knowledge to build a self-learning system with an example of the Rock-Paper-Scissors game.

8
Rock-Paper-Scissors – Intelligent Processing of Datasets

In order to present your results visually, you need a presentation. It is also desirable that this presentation is dynamic and lets you change the model parameters and see what the result will be. It is particularly important to have a cross-platform presentation, which will help to avoid many difficulties during its demonstration. In this chapter, with an example of the *Rock-Paper-Scissors* game, you will learn the following:

- How to develop an interface in Mathematica
- How to use Markov chains
- How to export this interface for the Web or for portable demonstration

Interface development in Mathematica

In this section, you will learn how to develop your own interface for a model demonstration using the Mathematica functions. To begin with, let's recall the rules of the Rock-Paper-Scissors game: two players independently show one of the three items, rock, scissors, or paper. The winner is determined as follows: scissors beat paper, paper beats stone, and stone beats scissors. Two identical items mean a draw. As you can see, if the players make their moves independently from each other and randomly, the average of their personal wins will be a draw. However, if the opponent has a strategy, then you can try to compute it.

The result of our development is the following interface:

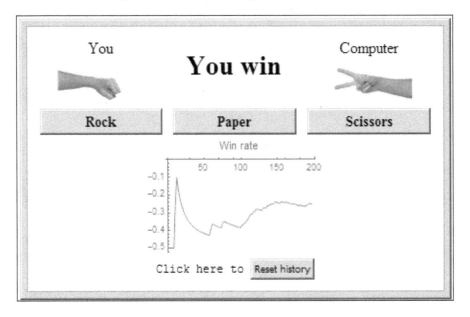

The interface consists of the following parts:

- Display of the current game's results: You can see your move and the computer's move, as well as the winner based on the rules.

- Buttons to select the next move: By clicking on one of the buttons, you make your move and the computer makes its own move simultaneously.

- Schedule of odds: You can track how the share of your wins changes in the course of time. Ideally, it should tend to constant.

- Button to reset the history: This is needed if you want to start from the beginning or change strategy.

Let's consider all the components in detail and see how to make this presentation interactive.

Interactivity in Mathematica is ensured with the help of the `Manipulate` function. The first parameter of this function is an expression that will be interactively calculated and the second is a set of parameters, a change that will affect the calculation. Here's how it will like for the function graph built according to the parameter changes:

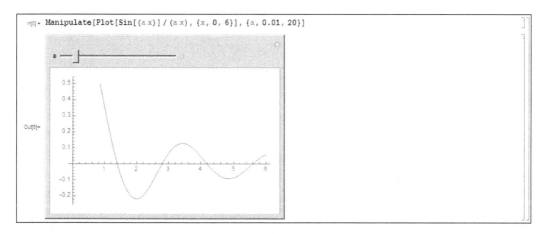

By changing the slider, we change the parameter value in the interval (0.01; 20), and the function graph is automatically redrawn. By clicking on the small **+** button near the slider, you can select a more accurate parameter value:

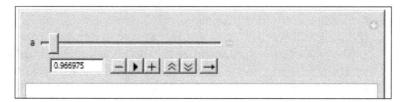

Let's define the required parameters for the interface of the Rock-Paper-Scissors game: `hist` — a history of moves, `yourTurn` — the current human's choice, `cTurn` — the current computer's choice, `hScr` — the number of wins by the human, `cScr` — the number of wins by the computer, `rates` — the history of the human's wins coefficient changes (for plotting), and `msg` — a message on the results of the current game. Since other functions will deal with the processing and display of these parameters, we are indicating that they should not be displayed on the screen (as it was with the slider in the previous example) using the `ControlType` and `AppearanceElements` parameters:

```
In[6]:= Manipulate[{},
    {{hist, {}}, ControlType → None},
    {{yourTurn, 0}, ControlType → None},
    {{cTurn, 0}, ControlType → None},
    {{hScr, 0}, ControlType → None},
    {{cScr, 0}, ControlType → None},
    {{rates, {}}, ControlType → None},
    {{msg, ""}, ControlType → None},
    AppearanceElements → None]

Out[6]=   {}
```

Thus, we have assigned initial values of the parameters while defining their type. To make an interactive model work, it is necessary to specify what should happen with the first launch. In our case, we need to define the functions used in future computations as follows: `winner` — the one who won the round based on the selection, `compMove` — a move the computer should choose (temporarily, we stick to a random strategy using the `RandomChoice` function. We'll consider more complex options in the following paragraphs). These functions are described in the initialization block as follows:

```
In[7]:= Manipulate[{},
        {{hist, {}}, ControlType → None},
        {{yourTurn, 0}, ControlType → None},
        {{cTurn, 0}, ControlType → None},
        {{hScr, 0}, ControlType → None},
        {{cScr, 0}, ControlType → None},
        {{rates, {}}, ControlType → None},
        {{msg, ""}, ControlType → None},
        AppearanceElements → None, Initialization :→
          ((* Test for who won a round*)
          winner[p1_, p2_] := Switch[Mod[p1 - p2, 3],
                            0, "Draw",
                            1, "Win",
                            2, "Lose"];
          (*Computer strategy*)
          compMove[data_] := RandomChoice[{1, 2, 3}];)]
```

In terms of initialization, we'll need two more functions responsible for the interface: dispButton — the function displaying the human's move selection buttons, which also adds data on moves to the history of the current game and dispText — the function that displays the choices of the human and the computer and the game's result. Before describing them, let's learn several Mathematica functions that will help us to build an interface. Firstly, these are the functions for tabulation, Row and Column:

```
In[8]:= Row[{1, 2, 3}, Frame → True]

Out[8]= 123

In[9]:= Column[{1, 2, 3}, Frame → True]

Out[9]= 1
        2
        3
```

To enable the user to make a choice, we need to create buttons for each item. This is done with the help of the `Button` function. This function has two parameters, the button text and the commands to be executed when it is clicked:

```
In[13]:= Row[{Button[#, Print["My choice is " #]]}] & /@
        {"Rock", "Paper", "Scissors"}

Out[13]= { Rock , Paper , Scissors }

        My choice is  Rock

        My choice is  Paper

        My choice is  Scissors
```

When the user clicks one of the buttons, a message is created with the user's choice. Thus, taking into account the logic of the game, the `dispButton` function will look like this:

```
dispButton[text_, move_] :=
Row[{Button[Text@Style[text, 16, Bold],
    cTurn = compMove[hist];
    yourTurn = move;
    msg = Switch[winner[yourTurn, cTurn],
        "Win", hScr++; "You win",
        "Lose", cScr++; "You lose",
        _, "Draw"];
    AppendTo[hist, {yourTurn, cTurn}];
    AppendTo[rates,
        hScr / Max[1, cScr + hScr] - 1 / 2];
    ]}];
```

Using the `Style` function, it is determined that the text will be displayed in a bold font with size 16. Pay attention to the `Switch` function. Depending on who wins the game, this function assigns the message about the win to the `msg` variable (which will be displayed in the main unit) and increases the number of wins by the computer or the human. The `AppendTo` function adds the current move to the history list. For clarity, we compute the coefficient of wins in such a way that it is greater than zero if a human has more wins than a computer.

The results' display function will be as follows:

```
dispText[play_, name_] :=
  Column[{Text[Style[name, 18]],
    Show[
      If[name === "You", ImageReflect[#, Left],
        #] &@ Switch[play, 1,        ,

        2,         , 3,         ,

        _, Graphics[{}]],
      ImageSize → {100, 46}]},
    Alignment → Center];
```

The `ImageReflect` function allows the mirroring of the picture; that's why, in order to show the move, it's enough to have only one set of images. In our game, we have the following numeric indication of the moves: 1 — stone, 2 — paper, 3 — scissors.

Now, let's turn to the interface main unit, which we had previously identified as an empty list, { }. With the help of the already described initialization block features, we can display such parts as the result of the game, move selection buttons, wins history graph, and the results reset button.

In order to make the presentation more dynamic, we'll use the `Deploy` and `Dynamic` functions.

Make sure that you have set the dynamic updates option. To do this, go to the **Evaluation** menu and make sure that the **Dynamic Updating Enabled** box is checked.

Using the `Deploy` function, we allow all the interactive attributes, such as buttons, to work but do not allow the form to be edited as a whole, for example, to change the size of images, their location, and so on.

With the help of the `Dynamic` function, we enable the form to be interactively updated, since we need to display the results of the game depending on the opponents' choices.

In general, the main computational block of the Manipulate function will look like this:

```
Manipulate[Deploy@Column[{Dynamic@
    Grid[{
      {(*Show outcomes*)
        dispText[yourTurn, "You"],
        Text@Style[msg, 30, Bold],
        dispText[cTurn, "Computer"]}, (*Play buttons*)
      MapIndexed[
        dispButton[#, #2[[1]]] &,
        {"Rock", "Paper", "Scissors"}],
      {(*Show win rate*)
        ListLinePlot[rates, ImageSize -> 200,
          PlotLabel -> "Win rate", AxesOrigin -> {0, 0}]
        , SpanFromLeft}
      }, ItemSize -> {{{10}}, Automatic}, Background -> White,
      Alignment -> Center, Spacings -> 1],
    (*Reset scores and play hist button*)
    Item[
     Row[{"Click here to ",
       Button["Reset history", hist = {{1, 1}, {2, 2}, {3, 3}};
         hScr = 0;
         cScr = 0;
         msg = "";
         cTurn = 0;
         rates = {};
         yourTurn = 0;]}], Alignment -> Center]
    }],
```

We have presented an interface in the form of one column that contains a table with the game results, selection buttons, and schedule of wins, as well as the string with the history reset button.

The Item function has allocated a separate row, where the statistics reset button will be located.

Markov chains

In our example, we did not use any strategy to maximize wins by the computer. However, let's imagine that a human does not just randomly choose his move, but his choice depends on his last move. Thus, we obtain a classical Markov chain with a finite state and discrete time. At every step, we can compute a new transition matrix based on the history of previous moves. Then, we choose the move that has the maximum probability, and we suppose that this will be a human's move. Based on this, the computer chooses a move to win against a human.

Let's consider step by step all the components of this approach. For a start, let's look at how the Markov chains are represented in Mathematica.

The `DiscreteMarkovProcess` function describes a time series whose elements constitute a discrete Markov chain. The first parameter of this function is the initial state of the chain, and the second parameter is the matrix of transition probabilities:

```
In[18]:= mChain = DiscreteMarkovProcess[{1, 0, 0},
            {{0, 1 / 3, 2 / 3}, {1 / 4, 1 / 4, 1 / 2},
            {1 / 2, 1 / 2, 0}}]

Out[18]= DiscreteMarkovProcess[{1, 0, 0},
            {{0, 1/3, 2/3}, {1/4, 1/4, 1/2}, {1/2, 1/2, 0}}]
```

Using the `MarkovProcessProperties` function, we can get all kinds of Markov process parameters, but we are interested in the transition matrix because it will be the basis to make predictions regarding the human's move:

```
In[19]:= MarkovProcessProperties[mChain, "TransitionMatrix"]

Out[19]= {{0, 1/3, 2/3}, {1/4, 1/4, 1/2}, {1/2, 1/2, 0}}
```

We should also mention an important capability, that is, the `Graph` function used to construct a graph of transitions. This graph is necessary for the visual presentation of Markov chain:

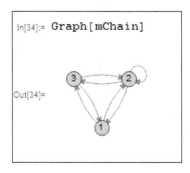

In order to find out what the transition matrix of the human's moves selection model will be, let's use the `EstimatedProcess` function, which is already known to us from *Chapter 5, Discovering the Advanced Capabilities of Time Series*. Let's consider an example of such an estimate process:

```
In[21]:= d = {{1, 1}, {2, 2}, {3, 3}, {1, 2}, {3, 1}};

In[22]:= est = EstimatedProcess[d[[All, 1]],
             DiscreteMarkovProcess[3]]

Out[22]= DiscreteMarkovProcess[{1, 0, 0},
             {{0, 1/2, 1/2}, {0, 0, 1}, {1, 0, 0}}]

In[23]:= matrix = MarkovProcessProperties[est,
             "TransitionMatrix"]

Out[23]= {{0, 1/2, 1/2}, {0, 0, 1}, {1, 0, 0}}
```

In order to predict a move of the human player, we need to select a row of the matrix with the number corresponding to the human's last move and select the column containing the most probable value. If there are several such probabilities, then we need to choose a column of data randomly. The selected value of the column will be the human's predicted next choice. We'll perform the selection of the maximum values using the `Reap` and `Sow` function. `Reap` gives the value of the expression together with all the expressions to which `Sow` has been applied to during its evaluation. A simple example will show how these work:

```
In[25]:= Reap[Sow[a1]; a2; Sow[a3]; Sow[a4]; a5; a6]

Out[25]= {a6, {{a1, a3, a4}}}
```

In order to select the array of the column indexes having the maximum probability, let's use the `Do` cycle function and the `If` selection function. Let's assume that the human's previous move was 1, that is, stone:

```
In[28]:= matrix

Out[28]= {{0, 1/2, 1/2}, {0, 0, 1}, {1, 0, 0}}

In[29]:= ind = Reap[Do[
            If[matrix[[1, i]] == Max[matrix[[1]]], Sow[i]],
            {i, 3}]][[2]]

Out[29]= {{2, 3}}

In[30]:= Flatten[ind]

Out[30]= {2, 3}
```

Thus, the human's next move may either be 2 or 3 with equal probability. You can choose one of these values using the `RandomChoice` function.

Pay attention to the `Flatten` function. It converts multilevel lists to a single level:

```
In[32]:= RandomChoice[Flatten[ind]]

Out[32]= 3
```

In general, the function of predicting the human's move based on the history of all the moves will look like this:

```
predictMove[data_] :=
  Block[
    {m = MarkovProcessProperties[
        EstimatedProcess[data[[All, 1]],
        DiscreteMarkovProcess[3]],
        "TransitionMatrix"]},
    RandomChoice[Flatten[Reap[Do[
        If[m[[Last[data][[1]], i]] ==
          Max[m[[Last[data][[1]]]]],
          Sow[i]],
        {i, 3}]][[2]]]]]];
```

If we assume that the next human's move will be 3 (scissors), then the computer's move should be 1 (stone). This means that the function of the computer's next move selection will be as follows:

```
compMove[data_] := Mod[predictMove[data] + 1, 3, 1];
```

The `Mod` function returns the remainder of the division. We also specify that the result should be with a shift of 1, since the indexes start from 1.

The final version of the initialization block will look like this:

```
Initialization :→

dispButton[text_, move_] := Row[{Button[Text@Style[text, 16, Bold],

    cTurn = compMove[hist];
    yourTurn = move;
    msg = Switch[winner[yourTurn, cTurn],
      "Win", hScr++; "You win",|
      "Lose", cScr++; "You lose",
      _, "Draw"];
    AppendTo[hist, {yourTurn, cTurn}];
    AppendTo[rates, hScr / Max[1, cScr + hScr] - 1 / 2];
  ]}];
dispText[play_, name_] :=
  Column[{Text[Style[name, 18]],

    Show[If[name === "You", ImageReflect[#, Left], #] &@

      Switch[play, 1,
```

```
                                , 2,                  , 3,                 ,

      _, Graphics[{}]], ImageSize → {100, 46}]}, Alignment → Center];
winner[p1_, p2_] := Switch[Mod[p1 - p2, 3],
                      0, "Draw",
                      1, "Win",
                      2, "Lose"];
predictMove[data_] :=
  Block[
    {m = MarkovProcessProperties[EstimatedProcess[data[[All, 1]],
        DiscreteMarkovProcess[3]], "TransitionMatrix"]},
    RandomChoice[Flatten[Reap[Do[
        If[m[[Last[data][[1]], i]] == Max[m[[Last[data][[1]]]]],
        Sow[i]],
        {i, 3}]][[2]]]]]];
compMove[data_] := Mod[predictMove[data] + 1, 3, 1];]]
```

Another approach, which is based on finding the most probable move of the opponent suggested by Jon Mcloone, can be read in the blog http://blog.wolfram.com/2014/01/20/how-to-win-at-rock-paper-scissors/.

Creating a portable demonstration

In order to make it possible to place this project on a web page or to do a cross-platform presentation on computers that do not have Mathematica installed, we need to use **Computable Document Format** (**CDF**). In this section, you will learn how to export the project to a file in this format.

To be able to view the presentation, you need to install the free software Wolfram CDF Player. Further information about the possibilities of the application and its distribution can be found at `http://www.wolfram.com/cdf/`.

The interactivity of the CDF file is fully ensured by the presence of the `Manipulate` function. Let's see how to create this file.

Let's select the final version of the Rock-Paper-Scissors game in our file. In the **File** menu, we need to select **CDF Export** and then **Standalone...** to create a cross-platform file or web-embeddable to embed the presentation into a web page. For example, let's select **Standalone**. After this, the following window appears:

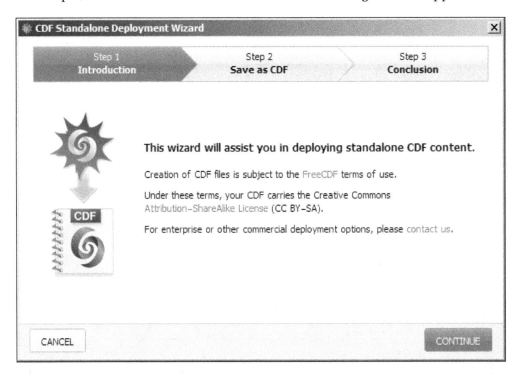

To go to further instructions, let's click on the **Continue** button. In response, the next dialog box appears:

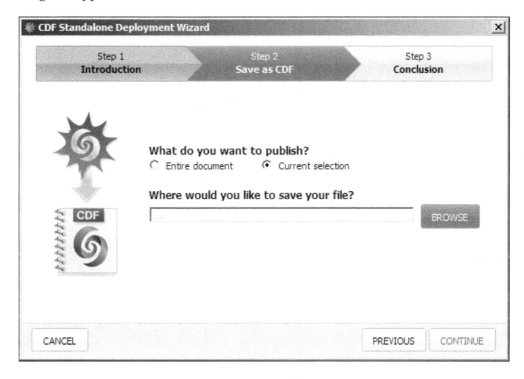

Here we need to state that we want to export the current selection and after this, choose the name of the file by clicking on the **Browse** button. Then, click on **Continue**:

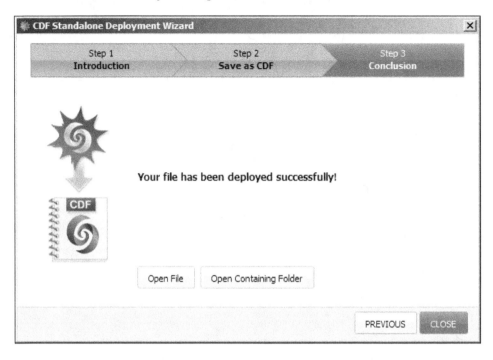

In response, we will get a message that the file has been successfully exported.

In case of exporting to the Web, we need to upload a file to the server and then it will provide an HTML code segment to be copied and pasted into an existing HTML file.

Thus, in order to demonstrate the results of your research, you need to copy the CDF file and the CDF player to your USB drive.

Summary

In this chapter, we learned how to create interactive forms to present research results with the help of functions such as `Manipulate` and `Deploy`. Continuing with the analysis tools, we have considered Markov chains and finding the transition probability matrix using the `DiscreteMarkovProcess` and `MarkovProcessProperties` functions.

In the end, we learned how to export the results to a file for cross-platform presentations with the help of the CDF Player.

Index

H

hypotheses
  mean, checking 93-100
  variance, checking 100-103
  on true sample distribution,
    checking 106-110

I

ImageCrop function 117
ImageDimensions function 117
ImageInstanceQ function 64
ImagePad function 117
image processing 117-121
ImageReflect function 131
images
  recognizing 63, 64
IndependenceTest function 103
information depository 78, 80
Inpaint function 119
interface
  components 126
  developing, in Mathematica 125-132
  implementation, performing with C/C++
    program 38-40
invertibility
  checking 88, 89

J

Java
  interacting with 49-52

K

kernel 10

L

LinearRegression method 113
LocationEquivalenceTest function 99, 100
LocationTest function 94

M

Manipulate function 127
Markov chains 133-137
Mathematica
  about 1
  brackets 12
  calling, from C 40-45
  cleaning functions 20
  data conversion 20
  data, importing 16
  data, importing from notebooks 26
  data types, URL 20
  expressions writing, main features 11-14
  front end 10
  information depository 77
  interface development 125
  kernel 10, 11
  system installation 1-7
  system, setting up 8, 9
  times series 74
  URL 2
mcc utility
  URL 39
mean
  hypotheses, checking 93-100
moving average model 80, 81

N

NearestNeighbors method 113
NET programs
  interacting with 45-48
  URL 49
NeuralNetwork method 113
notebooks
  data, importing 26, 27

O

observed data
  autocorrelation check 89
  dealing with 88
  invertibility check 89
  stationarity check 88

P

paclet 52
permissible data format
  Compression and Archive Formats 16
  Database Formats 16
  Data Interchange Formats 15
  for import 15
  Spreadsheet Formats 15
  Tabular Text Formats 15
  URL 16
  XML/HTML Formats 16
portable demonstration
  creating 138-140
predicting 112-116
probability automaton modeling
     method 122-124
process models, times series
  about 80
  autoregression model - moving average
     (ARMA) 82
  autoregressive process (AR) 81
  moving average model 80, 81
  selecting 84-88

R

R
  interacting, URL 54
  interacting with 52, 53
RandomChoice function 128
RandomForest method 113
RandomVariate function 94
RegularExpression function
  URL 22
  using 22, 23

S

sample dependence
  degree, checking 103-106
seasonal integrated autoregressive
     moving-average (SARIMA) 83, 84

SeedRandom function 103
SetAlphaChannel function 72
setHandler method 49
SmoothHistogram function 94
stationarity
  checking 88
strings
  importing 25
system
  installation 1-7
  setting up 8, 9

T

text information
  recognizing 68, 69
TextureSynthesis method 118
time series
  about 73, 80
  invertible process 89
  process models 80
  setting 74-77
  weakly stationary process 88
Transpose function 106

V

variance
  hypotheses, checking 100-103
VarianceEquivalenceTest function 102
VarianceTest function 101

W

WeatherData function 79
Wolfram CDF Player
  URL 138
Wolfram Symbolic Transfer Protocol
     (WSTP)
  about 35-38
  URL 45

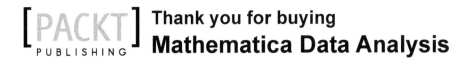
Thank you for buying
Mathematica Data Analysis

About Packt Publishing

Packt, pronounced 'packed', published its first book, *Mastering phpMyAdmin for Effective MySQL Management*, in April 2004, and subsequently continued to specialize in publishing highly focused books on specific technologies and solutions.

Our books and publications share the experiences of your fellow IT professionals in adapting and customizing today's systems, applications, and frameworks. Our solution-based books give you the knowledge and power to customize the software and technologies you're using to get the job done. Packt books are more specific and less general than the IT books you have seen in the past. Our unique business model allows us to bring you more focused information, giving you more of what you need to know, and less of what you don't.

Packt is a modern yet unique publishing company that focuses on producing quality, cutting-edge books for communities of developers, administrators, and newbies alike. For more information, please visit our website at www.packtpub.com.

Writing for Packt

We welcome all inquiries from people who are interested in authoring. Book proposals should be sent to author@packtpub.com. If your book idea is still at an early stage and you would like to discuss it first before writing a formal book proposal, then please contact us; one of our commissioning editors will get in touch with you.

We're not just looking for published authors; if you have strong technical skills but no writing experience, our experienced editors can help you develop a writing career, or simply get some additional reward for your expertise.

Mathematica Data Visualization

ISBN: 978-1-78328-299-9   Paperback: 146 pages

Create and prototype interactive data visualizations using Mathematica

1. Understand visualization functions used by scientists, engineers, and financial analysts.

2. Build a visualization system from scratch using low-level graphics primitives and interactive functionalities.

3. Learn how to visualize a wide range of datasets with the help of detailed explanations of code and theory.

R for Data Science

ISBN: 978-1-78439-086-0   Paperback: 364 pages

Learn and explore the fundamentals of data science with R

1. Familiarize yourself with R programming packages and learn how to utilize them effectively.

2. Learn how to detect different types of data mining sequences.

3. A step-by-step guide to understanding R scripts and the ramifications of your changes.

Please check **www.PacktPub.com** for information on our titles

Practical Data Analysis

ISBN: 978-1-78328-099-5   Paperback: 360 pages

Transform, model, and visualize your data through hands-on projects, developed in open source tools

1. Explore how to analyze your data in various innovative ways and turn them into insight.

2. Learn to use the D3.js visualization tool for exploratory data analysis.

3. Understand how to work with graphs and social data analysis.

4. Discover how to perform advanced query techniques and run MapReduce on MongoDB.

Python Data Visualization Cookbook

ISBN: 978-1-78216-336-7   Paperback: 280 pages

Over 60 recipes that will enable you to learn how to create attractive visualizations using Python's most popular libraries

1. Learn how to set up an optimal Python environment for data visualization.

2. Understand the topics such as importing data for visualization and formatting data for visualization.

3. Understand the underlying data and how to use the right visualizations.

Please check **www.PacktPub.com** for information on our titles

www.ingramcontent.com/pod-product-compliance
Lightning Source LLC
LaVergne TN
LVHW081344050326
832903LV00024B/1307